Long and Deep Tunnels: Integrated Design and Construction Approach

Long and Deep Tunnels: Integrated Design and Construction Approach

Editor

Neeraj Singh

scitus
academics

Long and Deep Tunnels: Integrated Design and Construction Approach
Edited by **Neeraj Singh**

ISBN: 978-1-68117-682-6
Library of Congress Control Number: 2016958834

SCITUS Academics LLC,
616, Corporate Way, Suite 2, 4766,
Valley Cottage, NY 10989

www.scitusacademics.com

Notice

Printed in United States of America on Acid Free Paper

Preface

A tunnel is an underground or underwater hallway, dug through the surrounding soil/earth/rock and enclosed except for entrance and exit, commonly at each end. A pipeline is not a tunnel, though some recent tunnels have used immersed tube construction techniques rather than traditional tunnel boring methods. A tunnel is relatively long and narrow; the length is often much greater than twice the diameter, although similar shorter excavations can be constructed, such as cross passages between tunnels. Anyone who sets out to construct underground works, finds himself having to tackle and solve a particularly complex civil engineering problem because, compared to surface constructions, it is extremely difficult to determine the basic design data of underground works in advance. It is not a question, as with surface construction, of gradually assembling materials with well known strength and deformation properties to build a structure which finds its future equilibrium in the desired final configuration, but of acting on a pre-existing equilibrium and proceeding in some way to a "planned disturbance" of it in conditions that are known only approximately. Another characteristic of underground works, well known to design and construction engineers, but not always given sufficient weight, is that very often the stage at which the structure is subject to most stress is not the final stage when the tunnel is finished and subject to external loads predicted at the design. The design and construction of long and deep tunnels, such as tunnels under mountains, signify a considerable challenge. *Long and Deep Tunnels: Integrated Design and Construction Approach* is not intended to teach how to design and construct such tunnels but to disclose the methods to identify the possible hazards associated to the process of designing and constructing long and deep tunnels.

The book is envisioned to practicing engineers, research practitioners, designers, advisers and students in mining, underground, tunneling, along with to foundation and geological engineers, urban planner and architects.

Table of Contents

1
CHAPTER

Urban Underground Space: Solving the Problems of Today's Cities

Wout Broere

Delft University of Technology, Geo-Engineering Section, Stevinweg 1, 2628 CN Delft, The Netherlands

HIGHLIGHTS

- Use of underground space in solving the problems of modern cities is discussed.
- Common infrastructure and environmental problems are paired with urban underground solutions.
- Awareness of engineers, planners and decision makers as an essential first step is discussed.

ABSTRACT

The world-wide trend of increased urbanisation creates problems for expanding and newly-developing cities alike. Population increase leads to an increased demand for reliable infrastructure, nowadays combined with a need for increased energy efficiency and a higher environmental awareness of the public. The use of underground space can help cities meet these increased demands while remaining compact, or find the space needed to include new functions in an existing city landscape. In many cases, underground solutions to urban problems are only considered if all other (above ground) options have been exhausted. When underground solutions are considered and evaluated from the planning or initial project stages onwards, more optimal solutions will become possible.

Use of the underground is not limited to large scale infrastructure projects. This paper also shows innovative use of the underground for commercial and residential use, storage, water conveyance and treatment, and heritage conservation, and highlights how use of underground can bring more optimal solutions for urban development.

Keywords: Underground space; Mass transit; Tunnelling; Urban planning

1. INTRODUCTION

The world is increasingly an urban environment. Since 2008 more than half of the world population lives in cities and the world population is expected to increase to roughly 10 billion people over the next four decades. As the world's rural population is projected to remain stable in this period, that increase will occur in urban areas. By 2050, 70% of all people will live in cities and the world urban population will have more than doubled compared to the turn of the century (UN, 2007 and UN, 2013).

In developing countries, where most of this uncontained population growth will take place, the rapidly expanding cities will need to meet the increased demands for infrastructure. Without efficient transport infrastructure, cities will sprawl away from the urban core, which strains the environment by creating more traffic congestion and travel time, loss of valuable farm land, and inequitable allocation of resources (Longman, 1998 and Chen, 2000). In the developed countries the urban expansion is less rapid, but the demographics of the population will change, with an increasingly large group over 60 (Angel et al., 2011). These population changes bring about new demands on the functions a city must provide and on the layout of the city, and call for continuous improvement in sustainable and resource efficient urban development (Camagni et al., 2002 and Jenks and Burgess, 2000).

Although high urban density can help cities become more energy and resource efficient, urban density alone is not sufficient to obtain a high standard of living. Comparing the most densely populated cities with the most liveable ones (Wikipedia, 2015 and Mercer, 2015) shows there must be other factors involved. This paper proposes that an efficient and integrated use of the underground is one of these factors and gives a brief overview of the possible solutions the underground offers to improve the factors contributing to quality of live: safety, health, convenience, and comfort (UN, 1961).

An urban population that is increasingly aware of the factors that improve quality of living, poses increased demands on their environment with respect to: reliable and safe transport of people and goods; dependable utilities, water distribution and sewerage systems; sustainability of the environment and limited urban sprawl; green spaces and recreational areas; reduced energy use and reduced emissions and noise levels; aesthetics and conservation of heritages; efficient use of real-estate and public space (Broere, 2012). In existing urban areas these demands pose significant challenges, as the space needed for developing new functions or relocating and improving existing ones is often not readily available. Placement of infrastructure and other facilities underground presents an opportunity to find the needed space, but it is often considered only as a last resort. This stems from a paradox, as the underground is considered only when surface space is exhausted and no other solutions exist any more for the complex urban problems to be tackled. This complexity and the obstructed access to the underground, created by the lack of space for the problem to be solved, almost inevitably lead to higher cost, giving underground solutions an expensive image, which in turn leads it to be considered less readily (Bosch, 2003 and Bobylev, 2009).

In newly developing metropolises that paradox need not exist, as initially the access to the underground is not obstructed and unique opportunities exist, if engineers, city planners and decision makers can come together and recognize that in order to reach an optimal solution, the underground option needs to be considered and used from the start (Sterling et al., 2012). The rapid and unconstrained expansion of these cities is only part of the problem. Structured city planning that includes underground options is still limited to a few examples, see for example Vähäaho, 2009 and Li et al., 2013. In general the awareness that the underground offers a possible solution for urban problems remains low, even though the issue has been raised on numerous occasions, e.g. Webster, 1914 and Sterling and Carmody, 1993. Recently Working Group 20 "Urban Problems – Underground Solutions" of the International Tunnelling and Underground Space Association (ITA) presented an overview of exemplary projects highlighting underground solutions to typical urban problems, in a renewed attempt to raise this awareness (Thewes et al., 2012). WG20 was founded in 2002 as part of ITA in order to identify urban problems that have been solved, or could have been solved, by the use of underground space or facilities, and to increase the awareness inside and outside the ITA. Other groups active in raising such awareness include the Committee on Underground Space (ITACUS), also part of ITA, and the Associated research Centers for Urban Underground Space (ACUUS), an international interdisciplinary association.

2. URBAN UNDERGROUND SOLUTIONS

Many dense urban environments face problems due to lacking infrastructure for transit, distribution of resources, goods and services. When paired with the demands listed above, these problems can be elaborated to include: traffic congestion; poor environmental conditions due to noise and air pollution; lack of safety, security, and protection against natural disasters and flooding; crowding and lack of space for work and recreation; restrictions when preserving aesthetic qualities and (cultural) heritages of the urban environment; aging infrastructure for distribution of resources, sewage conveyance and treatment; and combination effects of the above.

2.1. Traffic congestion

Probably the most recognized problem is the need for congestion relief in city streets. Time can be saved by using separated rail systems in order to reduce the rush hour traffic pressure. Hundreds of hours per worker per year can be saved in this way, as the cost of congestion in OECD countries is estimated to be equivalent to about 2 percent of the GDP (Godard, 2008).

But mass transit systems offer other benefits, as they tend to require less surface area than road traffic. Studies show that car traffic takes up 30 to 90 times more space than metro systems. Similarly, public road transport takes 3 to 12 times more space (Thewes et al., 2012). By moving from above ground car traffic to underground mass transit systems, enormous amounts of surface land can be freed up for other uses.

Continually improving tunnelling and excavation support technology adds to the success of urban rail systems. Advances in Tunnel Boring Machine (TBM) technology now allow tunnelling in more difficult ground conditions – even below the ground water table – with little disturbance to the surface. The surface influence is nowadays limited enough to realize bored tunnels even in highly sensitive city environments with protected cultural heritages, such as for example the historic city centres of Amsterdam and Rome (Burghignoli et al., 2013).

2.2. Pollution and noise

Highway noise and emissions from vehicles are recognized as pressing problems in urban areas. In order to reduce the noise impact, sound barriers may be erected, but the visual impact of such measures is major. It is often the case that residential property values near freeways are reduced due to high noise levels from cars and exhaust emissions. Also, there are associated health and safety issues for living close to a freeway.

Once again, moving passenger transport from cars to mass transit systems can reduce the noise and pollution impact at the local level, but also at a larger scale as mass transit systems tend to be more energy efficient and substantial energy savings can be obtained by the increased use of metro systems.

Alternatively, over the last few decades, many cities have constructed ring roads and roadway tunnels to improve their traffic conditions and to adapt the road network to the predicted demand. At the same time the travel times have improved and the impact of traffic on the surrounding residents has been reduced. Now, with city developments encroaching on existing ring roads, several cities have started to move surface sections of these ring roads below ground in order to further reduce their impact. Some examples are the double-deck tunnels in the A86 in Paris, the large diameter tunnels for the M30 in Madrid or the cut-and-cover tunnels for the A10 in Amsterdam (Samuel, 2006, Arnáiz and Bueno, 2009 and ZuidAs, 2015).

An even greater impact on their surroundings may be caused by the elevated highways, mainly constructed in the 1950s and 60s in a number of, mainly US, cities; for example Boston, Seattle, and San Francisco. These giant elevated structures through down-town areas are now seen as unsightly, noisy, possibly unsafe, and provide only limited access to areas adjacent to the freeway. Many cities are considering or in the process to replace the elevated highways by urban road tunnels. An example is the Alaskan Highway in Seattle, which when completed will be the largest diameter bored tunnel in the world (Gatti et al., 2013).

These transitions from surface or above-surface roads to underground solutions in more or less the same location are often complex and costly. While the original decision to build these roads above ground often focused only on direct construction costs, decision makers should include real-estate impacts, structural life span, and long-term sustainability when making such choices. This would help to avoid such unfavourable situations and reduce the life time cost of urban transport.

2.3. Protection against natural disasters

With concentration of population, urban areas are particularly vulnerable to failures in infrastructure due to ageing of the systems or those caused by other natural forces. Growth of population not only means more people are relying on the infrastructure, but at the same time that the man-made facilities may increase the severity of the disaster. For example, urbanization means more paved area leading to more severe flooding, as well as loss of water resources recharging groundwater.

Underground rivers can be constructed to increase run-off or divert storm water. Large diameter tunnels have been bored below cities such as Buenos Aires and Tokyo for this purpose (Dal Negro et al., 2012 and Miyao et al., 2000). The SMART tunnel in Kuala Lumpur takes this concept a step further, as this tunnel functions as a road tunnel during dry periods and is closed off for traffic and used as a storm-water tunnel during flooding periods (Abraham, 2008).

Also, it should be realised that the underground may provide a setting that is difficult to build in, but that underground structures offers better natural protection against environmental elements, including destructive weather and seismic events. Underground facilities and metro systems are less prone to earthquake damage have suffered little or no damage in major earthquakes (Wallis, 2010 and Tashiro and Mutou, 2013).

2.4. Lack of space and preservation of heritage and environment

Most of the underground examples above are not intended for a long-term human presence. This stems from the human preference to live, work and recreate above ground. Historically, underground structures were primarily intended for shelter or served as entry and connection points for mass transit systems. Over time, a wider range of functional facilities has taken up underground residence, but often still with a short intended stay for individuals below ground. Mostly the aim was to free surface space for other human needs and to improve the living conditions of cities. Examples such as underground car parks, shopping malls or underground storage facilities have been documented by Thewes et al. (2012).

Recently, the aim is more and more to not only keep surface space free and to create new space and functions, but to do so in a manner that preserves existing buildings and cultural heritages. This is especially true for public functions housed in historic monuments. A few examples of museums with newer underground extensions are the Louvre in Paris, the Rijksmuseum in Amsterdam and the Mauritshuis in The Hague. In all these cases additions to existing monuments have been realized without lowering the visual quality of the original buildings, and at the same time creating new floor space, with limited access points and a small footprint in order to preserve the security of the buildings and their collections.

And such underground extensions to monumental buildings are not limited to public buildings, but nowadays also include private residences, with the iceberg houses of London as the grander examples of the possibilities the underground offers (Reynolds, 2015).

2.5. Utilities and infrastructure

Focusing on the larger underground facilities it is easy to overlook the many utilities that are traditionally placed below ground. As discussed in Bosch and Broere (2011) the increasing amount of different utilities that is placed in the shallow subsurface strains the available space in the utility layer. Especially the addition of separated sewage systems for household waste water streams and storm water and of distributed heat-and-cold storage systems or shallow geothermal systems requires a large underground footprint if all placed directly in the ground. If not properly regulated and zoned, the increasing number of utilities creates underground space shortages in the shallow subsurface utility layer, and often causes increased surface disruptions given the increased number of parties that needs to inspect, repair or replace their underground utilities.

Utility tunnels, small tunnels placed at shallow depth that in turn contain cables and ducts for different utilities, are a solution here. Not only do they reduce the effective footprint for utilities, as they can be placed closer together inside the utility tunnel, but these utilities can also be inspected and repaired without the need to dig in the subsurface. Hunt et al. (2014) shows that already for a limited number of utilities such a multi-utility tunnels can be cost effective.

Not only utility pipelines, but also the associated treatment plants and facilities can be placed underground. Underground waste water treatment plants such as in Stockholm, Rotterdam or Guangzhou City show how such facilities not only free up space at surface, but also reduce the olfactory impact to neighbouring residences normally associated with these plants and help reduce the overall environmental impact of waste water (Watertechnology, 2015 and Waterworld, 2010).

3. CONCLUSIONS

Underground development is an important tool in developing and reshaping urban areas to meet the challenges of the future. Placement of infrastructure and other facilities underground presents an opportunity for realizing new functions in urban areas without destroying heritages or negatively impacting the surface environment, and at the same time brings opportunities for long-term improvements in the environmental impact of cities and more efficient use of space and resources. These benefits are there for existing, redeveloping cities, but can be implemented for newly developing cities more easily and more cost effectively, for even greater benefits.

The number of examples given in this paper is limited, but already in this small set it can be observed that many of these underground solutions can solve or help improve multiple of the problems that urban developments face: traffic congestion; environmental problems; lack of (green) space; need for protection against disasters; lack of infrastructure for food, energy, water and sanitation.

When planning and developing cities, the underground should not be overlooked by planners, engineers and decision makers. Raising the awareness of the benefits underground space can bring is a first step towards a systematic use of underground space in urban environments.

REFERENCES

1. Abraham, K., 2008. Smart tunnel the unique dual purpose solution for Kuala Lumpur. In: Enlightened Underground, Centrum Ondergronds Bouwen.
2. S. Angel, J. Parent, D.L. Civco, A. Blei, D. Potere **The dimensions of global urban expansion: estimates and projections for all countries, 2000–2050.** Prog. Plan., 75 (2011), pp. 53–107
3. M. Arnáiz, P. Bueno **Madrid: a global example of the development of safe and efficient underground urban road and rail infrastructures** Mod. Tunnell. Technol., 4 (2009), p. 002
4. N. Bobylev **Mainstreaming sustainable development into a city's master plan: a case of urban underground space use** Land Use Policy, 26 (2009), pp. 1128–1137
5. Bosch, J., Broere, W., 2011. The urgent need for physical planning of the underground in the Netherlands. In: Särkkä, P. (Ed.), Proceedings WTC2011 Underground Spaces in the Service of a Sustainable Society, pp. 150–151.

6. J.W. Bosch **De paradox van ondergronds bouwen; meervoudig ≠ eenvoudig** Delft University Press (2003)

7. Broere, W., 2012. Urban problems – underground solutions. In: Zhou, Y., Cai, J., Sterling, R. (Eds.), ACUUS 2012 Advances in Underground Space Development, pp. 1–12.

8. Burghignoli, A., Callisto, L., Rampello, S., Soccodato, F., Viggiani, G., 2013. The crossing of the historical city centre of Rome by the new underground line C: a study of soil-structure interaction for historical buildings. In: Bilotta, E., Flora, A., Lirer, S., Viggiani, C. (Eds.), Geotechnics and Heritage.

9. R. Camagni, M.C. Gibelli, P. Rigamonti **Urban mobility and urban form: the social and environmental costs of different patterns of urban expansion** Ecol. Econ., 40 (2002), pp. 199–216

10. D.D. Chen **The science of smart growth** Scient. Am., 283 (2000), pp. 84–91

11. Dal Negro, E., Boscaro, A., Laham, D., 2012. Successful TBM operations in urban environment: the "Arroyo Maldonado" project in Buenos Aires. In: Proceedings of the ITA-AITES World Tunnel Congress 2012 – Tunnelling and Underground Space for a Global Society.

12. U.C. Gatti, G.C. Migliaccio, L. Laird **Design management in design-build megaprojects: SR 99 bored tunnel case study** Pract. Period. Struct. Des. Construct. (2013)

13. Godard, J., 2008. Should we/can we avoid underground urban mass transit systems? In: Proceedings of the ITA-AITES World Tunnel Congress 2008, Agra, India.

14. D. Hunt, D. Nash, C. Rogers **Sustainable utility placement via multi-utility tunnels** Tunnell. Undergr. Space Technol., 39 (2014), pp. 15–26

15. M. Jenks, R. Burgess **Compact Cities: Sustainable Urban Forms for Developing Countries** Taylor & Francis (2000)

16. H.Q. Li, A. Parriaux, P. Thalmann, X.Z. Li **An integrated planning concept for the emerging underground urbanism: deep city method part 1 concept, process and application** Tunnell. Undergr. Space Technol., 38 (2013), pp. 559–568

17. Longman, P.J., 1998. Who Pays for Sprawl, Hidden Subsidies Fuel the Growth of the Suburban Fringe. US News and World Report 27.

18. Mercer, 2015. Mercer 2015 Quality of Living. (retrieved on 06.07.15).

19. H. Miyao, Y. Kamoshita, M. Kanai, K. Fukumoto **Underground river tunnel for flood control by shield tunneling** The Tenth International Offshore and Polar Engineering Conference, International Society of Offshore and Polar Engineers (2000)

20. Reynolds, E., 2015. NY-Lon underground. In: Admiraal, H. (Ed.), Think Deep: Planning, Development and Use of Underground Space in Cities. ITACUS – ISOCARP.

21. P. Samuel **Innovative roadway design** Policy Study (2006)

22. R. Sterling, H. Admiraal, N. Bobylev, H. Parker, J.P. Godard, I. Vähäaho, C.D. Rogers, X. Shi, T. Hanamura **Sustainability issues for underground space in urban areas** Proc. ICE-Urban Des. Plan., 165 (2012), pp. 241–254

23. R. Sterling, J. Carmody **Underground Space Design** Van Nostrand Reinhold Company (1993)

24. Tashiro, Y., Mutou, Y., 2013. Disaster-prevention measures for Tokyo metro tunnels. In: Anagnostou, G., Ehrbar, H. (Eds.), World Tunnel Congress 2013 Geneva – Underground – the way to the future!, pp. 275–282.

25. Thewes, M., Godard, J., Kocsonya, P., Nisji, J., Arends, G., Broere, W., Elioff, A., Parker, H., Sterling, R. (Eds.), 2012. Report on Underground Solutions for Urban Problems, ITA Working Group Urban Problems – Underground Solutions. 11, ITA-AITES.

26. UN, 1961. An Interim Report on the International Definition and Measurement of Levels of Living. Technical Report Ecn.3/270/Rev.1-E.CN.5/353, United Nations (UN).

27. UN, 2007. World Population Prospects: The 2007 Revision. Technical Report. United Nations, Department of Economic and Social Affairs, Population Division.

28. UN, 2013. World Population Prospects: The 2012 Revision. Technical Report ESA/P/WP.228. United Nations, Department of Economic and Social Affairs, Population Division.

29. Vähäaho **Underground master plan of Helsinki, a city growing inside bedrock** P. Kocsonya (Ed.), ITA-AITES World Tunnel Congress 2009 Safe Tunnelling For The City and Environment, Hungarian Tunneling Association (2009)

30. Wallis, S., 2010. Santiago Metro Withstands Massive Earthquake..

31. Watertechnology, 2015. Henriksdal Wastewater Treatment Plant, Stockholm, Sweden. (retrieved on 08.07.015).

32. Waterworld, 2010. Largest Underground MBR Plant in Asia Completed.

33. G.S. Webster **Subterranean street planning** Ann. Am. Acad. Polit. Soc. Sci., 51 (1914), pp. 200–207

34. Wikipedia, 2015. List of Cities by Population Density. (retrieved on 06.07.15).

35. ZuidAs, 2015. Amsterdam ZuidAs-Dok. (retrieved on 06.07.15).

2
CHAPTER

Underground Space as an Urban Indicator: Measuring use of Subsurface

Nikolai Bobylev, [1],

Institute of Earth Sciences, Saint Petersburg State University, 7-9, Universitetskaya nab., Saint Petersburg 199034, Russia

Saint Petersburg Research Centre for Ecological Safety of the Russian Academy of Sciences, 18, Korpusnja ul., Saint Petersburg 197110, Russia

Peter the Great Saint-Petersburg Polytechnic University, Polytechnicheskaya, 29, St.Petersburg, 195251, Russia

HIGHLIGHTS

- Urban underground space ought to be an important component of urban indicator lists.
- Three key quantitative indicators suggested to monitor urban underground space use.
- Average urban underground space use density in cities is up to about 0.05 m^3/m^2.
- Developed underground space volume per person in cities is up to about 10 m^3.

ABSTRACT

Use of Urban Underground Space (UUS) has been growing significantly in the world's biggest and wealthiest cities. UUS has been long acknowledged to be important to the urban development agenda: sustainability, resilience, livability, and creating a better urban environment in particular. These issues are traditionally monitored using urban indicators, however UUS has not been properly included and considered in urban indicator lists (sets or systems) yet – the gap this paper is aiming to bridge. The paper reviews existing approaches to the composition of urban indicator lists, highlighting indicator types, challenges related to data collection, and agencies that are concerned with the issue. Further the paper has identified the importance of UUS inclusion in the lists that give integrated assessment and monitor urban sustainability, resilience, climate change adaptation and mitigation, as well as progress towards smart, livable, and compact cities. Existing global quantitative data on UUS have been examined in 8 cities; and three key indicators (descriptors) were suggested to monitor UUS use: Developed UUS volume (m^3); UUS use density (m^3/m^2); and Developed UUS volume per person (m^3/person). Current average UUS use densities in cities are identified as up to about 0.05 (m^3/m^2) (which can be interpreted as a virtual depth of UUS use of 5 cm), and the developed UUS volume per person is up to about 10 m^3/person; while city central areas (central business districts) can have a virtual depth of developed UUS of several metres (m^3/m^2). Compatibility, comparability, uniformity, and sustained monitoring of urban indicators data (including UUS indicators) found to be posing significant challenges to the research across geographies, and industry/economic sectors.

Keywords: Urban underground space; Urban indicators; Urbanisation; Sustainability; Resilience; Density

1. INTRODUCTION

Urban Underground Space (UUS) use has been growing significantly in the world's biggest and wealthiest cities. Arguably, the main driving factors of this growth were lack of surface space and a need for a better environment, including abatement of motor traffic and pollution problems. Generalising, we can suggest that awareness of *the urban sustainability agenda* and a need to make cities more liveable have been growing concurrently with intensification of UUS development.

Indeed, UUS development can contribute a lot to urban sustainability, ranging from local renewable energy provisioning to urban space cohesiveness and aesthetics. Sustainability issues related to UUS use were raised by Carmody and Sterling, 1993,Sterling, 1997, Bobylev, 2006, Bobylev, 2011, Rogers, 2009 and ITACUS, 2010, and systematised by Sterling et al. (2012).

Measuring sustainability is an important subject, both in scholarly terms and as a policy informing tool. Lists of urban indicators or urban sustainability indicators have been adopted by many cities, countries, and international organizations to monitor progress in sustainable urban development. Sustainability is just one of the concepts that require to be informed by urban indicators; most recently the concepts of *ecosystem services,resilience, smart cities* have been developed and require input of urban data. Thus urban indicators become a more general notion, pertaining to developing, collecting, and analysing data from different aspects of urban life and then applying this knowledge to develop a better urban environment.

Usually urban indicators are presented in a form of lists where individual indicators are grouped according to a subject or knowledge area. Data behind these lists have different degrees of comprehensiveness and accuracy in terms of indicator monitoring. Specific indicators can have a variety of methodologies of data collection, ranging from field monitoring and comprehensive numerical data to expert estimations and rankings.

In spite of acknowledgement of UUS importance to the concepts and urban issues highlighted by use of urban indicators (e.g. sustainability, resilience), this subject has not made it yet into routine urban indicator lists. The importance of UUS as an urban activity sector is on a par with long established urban sectors as transport (widely used indicator: motorisation rate), land use and planning (widely used indicator: built stock density), environment (widely used indicators: air pollution, water quality).

The undeservingly marginal role of UUS in urban sustainability and resilience discourse is reflected by the fact that the UUS topic has not made it yet into executive summaries of the most known *policy documents related to urban development*, i.e. United Nations Human Settlements Programme State of Cities Reports (UN Habitat, 2006 and UN Habitat, 2013a); United Nations Environment Programme Geo Outlook (UNEP, 2012); The World Bank Annual Reports and Urbanization Reviews (World Bank, 2012); Organisation for Economic Co-operation and Development Infrastructure Outlooks (OECD, 2006 and OECD, 2008). However, the progress regarding mainstreaming UUS into urban agendas has been made. The United Nations Secretary General's formal address to the International Tunnelling Association conference in Bangkok in 2006 highlighted UUS relevance to global development and urban sustainability agendas (UN, 2012b). Famous architect Norman Foster highlighted the strategic importance of UUS as well: "One of the greatest challenges facing mankind is to achieve higher density while at the same time improving urban existence. The underground has enormous potential for realising spatial benefits" (Foster, 2011).

This paper argues that considering UUS in urban indicators lists will help to better understand the role that UUS plays in urban sustainability, resilience, and creating a better urban environment and life in general. Considering underground space as an urban indicator will help both: (1) better urban policy informing, and (2) better understanding of UUS sector industries needs and directions for development.

One challenge is arguing and promoting UUS inclusion into urban indicators; another one is to suggest how to do it. The paper will review existing approaches to composition of urban indicator lists, highlighting indicators types, problems related to data collection, and agencies that are concerned with the issue. Further the paper will examine existing data on UUS globally, trying to make sense of what actually can be measured in UUS and how this could help to better inform sustainability and resilience agendas. Finally, some possible UUS related indicators and their descriptors will be suggested, along with available data cross-sector analysis and comparisons.

2. URBAN INDICATORS

2.1. Emergence, systems, agencies

The emergence of *the urban indicator theme* stems from *the Sustainable Development Concept* (Brundtland Commission, 1987), and one of the first widely accepted set of indicators was part of the Local Agenda 21 (United Nations, 1995). Attempts for a singular comprehensive indicator set were made by United Nations Sustainable Development office (1998), comprising the list of 134 sustainability indicators. During about two decades of urban indicator research numerous lists, sets or systems of indicators have emerged. These lists were adopted by a variety of agencies and at a variety of levels (from national to local), which suggests the importance of diversity and fine turning of indicator lists. The need for development and structuring of urban indicators in a specific context was reflected in recent scholarly publications dealing with regionalization (e.g. Gonzalo et al., 2015, Michael et al., 2014 and Shen et al., 2011), and the development of different indicator tools that aim to analyse urban sustainability (e.g.Castanheira and Bragança, 2014). Some questions regarding

UUS and indicators include whether UUS should be featured in any specific lists (i.e. pertaining to a certain level or developed for any specific purpose), and/or if it is appropriate to have UUS in any lists dealing with topics of UUS concern: e.g. sustainability.

Urban indicator lists, sets, or systems have been developed by different *agencies*. The most famous of them aim on comprehensiveness and global applicability United Nations (2007), Organisation for Economic Co-operation and Development (OECD, 2004), the World Bank (World Bank, 2015), European Union (Eurostat, 2009), World Health Organisation (WHO, 1999), United Nations Human Settlements Programme (UN-Habitat, 2013b).

There is a number of specific *assessment tools*, that are in fact using urban indicators, as summarised and classified by Gonzalo et al. (2015), who considered 13 systems. Amongst them are certification systems developed for urban related industries: construction, planning, transportation. Major international systems are Leadership in Energy and Environmental Design (LEED) (US GBC, 2009), Building Research Establishment Environmental Assessment Method (BREEAM) (BRE Global, 2011), Sustainable Community Rating (SCR, 2015), Key Indicators for Territorial Cohesion and Spatial Planning (Daly and González, 2013).

A number of a *large scale research projects* were aimed at creating a comprehensive online databases of urban indicators: European Common Indicators (European Commission, 2003), Urban Audit (Urban Audit, 2004), European Thematic Network on Construction and City Related Sustainability Indicators (CRISP Project, 1999), Cities Environment Reports on the Internet (CEROI Project, 2010). Unfortunately, in majority of cases, the data update has been discontinued after the projects have ended, nonetheless, these projects remain an important methodological reference.

2.2. Types and classifications

The most traditional approach to create an indicator list would be to group indicators according *to three pillars of sustainability* (environment, economy, society). However nowadays a purpose-driven approach prevails in most urban indicator lists, i.e. broad indicator categories reflect agenda or concerns of the list proponent. Table 1 exemplifies aggregated indicator categories of the highest hierarchical level presented by several agencies.

Table 1. Examples of major aggregated categories of urban indicators as suggested by different proponents.

Name and purpose of indicator system/list	Urban indicators reflecting United Nations agenda and the progress towards millennium development goals	Indicator system to assess urban sustainability for the Spanish context	A system of indicators for the city of Thessaloniki, Greece
Major aggregated categories	• Shelter • Social development and eradication of poverty • Environmental management • Economic development • Governance	• Site and soil • Urban morphology • Mobility and transport • Nature and biodiversity • Building and housing • Energy • Water • Materials • Waste • Pollution • Social aspect • Economic aspect • Management and institution • Innovation	• Economy and population • Land and urban planning • Energy • Transportation • Agriculture, livestock, fishery • Industry • Tourism • Air pollution and climate change • Water resources and sea environment • Solid waste • Biodiversity • Health • Education, research, and technology
Reference	United Nations Human Settlements Programme (2013b)	Gonzalo et al. (2015)	Moussiopoulos et al. (2010)

As Table 1 reflects, indicator lists tend to be as comprehensive as possible, prioritising main concerns of the developer (e.g. note category "poverty" in the UN Habitat list). Urban indicator lists presented in Table 1 represent *different scales* – from global to national and a city one. Indicator assessment is done at an urban (city) level in any system, but the UN Habitat list is concerned with global relevance, while the Thessaloniki list is concerned just with the issues relevant to this particular city. Urban indicators bring*different meaning to different levels* (Lynch et al., 2011). At a local level the indicators are mainly used to monitor and inform urban development by city authorities; at the regional and national levels indicators inform development programmes and policies; at the global level the indicators are used to inform policies of international development agencies, including setting cross sector priorities (e.g. financing, climate change) that go beyond urban agendas.

Indicators differ in actual *approaches to measure them*. Significant division is between*quantitative and qualitative indicators*. This division can be referred to as different measurement methods, or different descriptors. An indicator formulation usually reflects what we want to know according to our (e.g. sustainability) goals e.g. "outdoor air quality"; the descriptor would reflect on available data we can monitor, e.g. "proportion of population exposed to SO_x above x mg/m^3", or "$PM_{2.5}$ mean annual exposure, % of population exceeding World Health Organisation guidelines level" (World Bank, 2015). Similar quantitative indicators in different indicator lists can have different descriptors.

Descriptors can differ in data collection methods, which could make comparisons amongst different indicator systems difficult (Bobylev, 2009b, Bobylev, 2010a and World Bank, 2011). Data on qualitative indicators is presented in a form of expert judgements, an example of this indicator type is "effectiveness of environmental planning guidelines".

Proxy indicators, or proxy descriptors are used when we assume a correlation between different processes, e.g. daytime population density can be measured as density of mobile phones registration in a network cell. Proxy descriptors are important in public transport, land use, energy, and many other indicators.

2.3. Challenges

Persisting challenges of urban indicators research and practice are in *comprehensive data collection and handling*. As it was reflected in the previous sections: (1) actual indicator data would be collected by different agencies (industries); (2) a variety of methodologies for data collection are used; (3) data have different degrees of reliability.*Monitoring* urban environment and sustainability is a holistic task, however urban indicators data required for this task are mostly specific to the relevant industry. Data on urban indicators are usually compiled by local authorities' office, but data on specific indicators are collected in the form of statistics which are provided by industries (e.g. water related data would be provided by a water management company). Usually industries would have their own statistics and other types of monitoring indicators (e.g. key performance indicators) that are focused on their needs. This means that there is often *a mismatch between data available from industry and data required for urban indicators*. For instance, one of the urban sustainability indicators is "proportion of rainwater reused", but a water management company would usually know just volumes of water treated by the sewerage system. If over the years the latter decline (when adjusted for population and other factors) one can assume that more rainwater is reused. This example illustrates a challenge of actually feeding and interpreting data in urban indicator lists.

Compatibility and comparability of urban indicator data are important issues as well. Making comparisons between cities using indicators is an essential base for policy advice and judging on progress towards sustainability. The significance of analysis across geographies and indicator systems has been acknowledged by many, including the World Bank. The World Bank, which has been doing comprehensive research on the subject, admits that even its own urban indicator data "should be used with care, and because of different data collection methods and definitions may have been used, comparisons can be misleading" (World Bank, 2015). This can be illustrated on such a trivial example as "urban area", which can refer to (1) administrative district, (2) build up area of a certain density, (3) it can include or exclude large water bodies and green parks.

Table 2 presents a summary of discussed issues in urban indicators research, giving a UUS perspective on the subject.

Table 2. Summary of major contemporary issues regarding urban indicators.

An issue or a challenge	State-of-the-art	Urban Underground Space (UUS) perspective
Purpose	Policy driven; urban sustainability, resilience agenda; can be fine-turned for specific agency or industry needs; can be relevant to global environment/development agenda	(1) To provide more comprehensive perspective on urban issues (2) To help UUS being developed in a more sustainable way
Level/scale	Urban data is used for urban, regional, national, or global information	All scales are relevant to UUS, for details see Table 9
Geographies	Cities in different regional groups may have specific to them urban indicator needs	Ground conditions (hard rock or weak soils) would be the main diversifying factor for UUS
Agency/industry	Indicators may reflect specific industry needs (e.g. transport focused), or contributing specific industry data to holistic urban policies	(1) UUS is relevant to holistic urban policies (e.g. energy, environment) (2) UUS agenda can be combined with agendas of close (sister) industries (e.g. water management, transport, waste) (3) UUS has its own distinct sustainability, resilience, climate change agendas
Types	Qualitative, quantitative, proxy	For suggested types of UUS indicators see Table 5
Continuity	Some indicator systems have been feeding data in for some time, but have major gaps or monitoring have been discontinued	Not relevant yet, UUS is just emerging as an indicator
Compatibility and comparability	Variety of methodologies for data collection make challenging analysis across indicator systems	UUS data is collected in different units (e.g. floor area or volume), accuracy can suffer while to making data compatible
Data quality	Variety quality and accuracy of data requires making tolerances in analysis and conclusions	As UUS is an emerging indicator, data is scarce yet, comparisons allow for big tolerances

3. Urban underground space as an urban indicator

3.1. Background and relevance

Urban Underground Space (UUS) can be defined as a geo space beneath urban areas, including wider areas of UUS that provide direct services to a city, like groundwater supply or geothermal energy. UUS encompasses geologically formed rocks and soils, and artificial spaces, as well as caverns of various origins.

Manmade UUS includes *Urban Underground Infrastructure*, which can be defined as a set of artificial structures, located entirely or partially below ground level, interconnected physically or functionally (Bobylev, 2007). UUI is represented by a variety of utilities, rail and motor tunnels, buildings' basements used as storages, garages, public pedestrian and shopping zones, etc.

UUS services, or functions in the city, can be summarised as storage (e.g. food, water, oil, industrial goods, waste); industry (e.g. manufacturing); energy production (e.g. geothermal energy sourcing); transport (e.g. railways, roads, pedestrian tunnels); utility supply (e.g. water, gas, electricity and communications) and waste disposal (e.g. waste water); and provision of public space (e.g. shopping centres, hospitals, parking, civil defence structures) and private space (e.g. cellars, houschold garages and dwellings) (amended after Bobylev, 2009a and Sterling et al., 2012).

As it was mentioned in the introduction, urban indicators help to monitor cities' performance and progress towards sustainability, resilience, climate protection, liveability, etc. UUS is very relevant to all this issues, as it was shown during the last decade of UUS research. Table 3 names urbanisation related scholarly concepts, and explains UUS relevance to them, giving key references to a research on the subject.

Table 3. Global development and urbanisation related concepts and Urban Underground Space (UUS).

A concept and reference to the Urban Underground Space (UUS) research	Summary of major Urban Underground Space (UUS) relevant issues
Sustainability (Sterling et al., 2012)	Rational use of UUS resources Rational land use Combating urban sprawl and compact city Geothermal energy (deep) and shallow subsurface heat exchange) Urban infrastructure efficiency (transport, water, others)
Resilience (Sterling and Nelson, 2013; Bobylev et al., 2013; Makana et al., 2016)	Urban natural and artificial disasters preparedness Emergency response and civil defence facilities Mitigation of city scale adverse environmental impacts (e.g. urban heat island effect) Critical infrastructure reliability
Climate change adaptation and mitigation (Bobylev, 2009b, 2013)	Urban networks energy efficiency (mitigation) Stable temperature mode benefits while locating urban functions underground (mitigation) Enabling urban compactness (mitigation) Underground infrastructure facilities for urban climate change adaptation Adaptation of urban underground infrastructure to climate change (reflecting changes in water balance, extreme temperatures)
Smart city (Bobylev, 2014)	Greater use of information and communication technologies to enable more efficient use of existing urban underground infrastructure facilities (e.g. water sewers)
Liveable city (Hunt et al., 2016)	Compact and high quality public spaces; Enhancing urban green and recreational areas by putting infrastructure underground
Compact city (Bobylev, 2009a; Wende et al., 2010)	Densification Quality of life and the environment Proximity
"0-land use" (Vahaaho, 2013)	A concept of "0-land use" is an idealistic approach to urban growth and development using just underground space. The concept originates from the city of Helsinki, Finland, where significant advancements in UUS planning has been made

As Table 3 shows, inclusion of UUS into urban indicators is quite justified by the number and significance of the issues related to underground space.

3.2. Urban underground space in urban indicator lists

Urban indicator systems have been unfair to UUS and urban underground infrastructure so far in comparison to other sectors and infrastructures, e.g. transport, water, telecom.Table 4 illustrates this by showing the number of indicators of the lowest level in respective groups (e.g. transport) in different indicator lists. Table 4 suggests where UUS indicators might be appropriate in the indicator lists. Table 4 is build upon three indicator lists, which were earlier discussed in Table 1. The aim is to show needs and opportunities for UUS indicators to be included in the indicators list of *variety of purposes, levels/scales, and originating agencies*. Readers can see full lists (several pages each) in the respective publications. As it was cited in the previous chapters, there is not and should not be just one universal, fit for all purposes, urban indicator list. Similarly, *there should not be only one way or an indicator reflecting UUS related issues*, but rather UUS indicators should not be forgotten in the lists.

Table 5 was developed on the basis of the "*proposed UUS indicators at lowest level*" section in Table 4, and further suggests specifics of UUS indicators. These UUS indicators were developed on the basis of analysis of urban indicator lists and UUS relevance to the goals of these lists. Arguably, UUS can be included into urban indicator lists as a separate category (such as transport), however (1) since UUS is relevant to many categories (see Table 4), and (2) UUS development is not a goal in itself, it is deemed more beneficial to have indicators reflecting UUS use included into traditional categories (as for example shown in Table 4).

Table 4. Suggestions for Urban Underground Space (UUS) indicators inclusion into urban indicator lists.

Major aggregated category	Indicator system – level/scale (i.e. "Global" indicates Global relevance)	Existing indicators at the lowest level (only relevant are shown)	Proposed Urban Underground Space (UUS) indicators at the lowest level
Environmental management	Global	Planned settlements	UUS integrated into master plan
Environmental management	Global	Disaster prevention and mitigation instruments	Adequate consideration and use of UUS in disasters preparedness
Site and soil	Regional	Planning according to climatic zones	Planning according to UUS conditions (soils, groundwater)
Urban morphology	Regional	Design and quality of public space	Design and quality of public space, including underground public spaces
Building and housing	Regional	Proportion of abandoned or unused buildings	Underground structures that are abandoned or inefficiently used
Building and housing	Regional	Consideration of solar orientation in building design	Consideration of UUS use (also adjacent) in building design
Energy	Regional	Proportion of buildings whose energy rating is higher than average	Proportion of energy efficient UUS premises
Energy	Regional	Proportion of self-sufficiency with renewable energy	Proportion of buildings using UUS for energy sourcing or storage
Innovation	Regional	Innovation practices based on BRE Global (BRE, 2011)	Innovations in infrastructure and UUS use
Economy and population	City	Population density	UUS use density
Land and urban planning	City	Share of land use by activity (function)	UUS use by function
Energy	City	Renewable energy	Share of renewable energy by source, including underground
Transportation	City	Public transportation vehicles average speed	Average speed of passenger travel on underground public transport
Industry	City	Share of industries with environmental management systems	Share of urban functions that are underground and not consuming urban land
Land use	*Additional suggestion (indicator was not mentioned in the reviewed systems)	Share of land cover by cover type (asphalt, soil, building, etc.)	Stratification of UUS use by depth

Table 5. Urban Underground Space (UUS) indicators for urban indicator lists.

Urban Underground Space (UUS) indicator proposal (lowest level of hierarchy)	Possible types of indicator and explanations. Suggestions for units	
	Qualitative	Quantitative
UUS integration into master plan	Yes/no; Quality and detail rating	
Adequate consideration and use of UUS in disasters preparedness	Yes/no; Quality and detail rating	
Planning according to UUS conditions (soils, groundwater)	Yes/no	
Design and quality of public space, including underground public spaces	Expert rating [important is to have a holistic approach to public/open spaces, including large underground developments]	
Underground structures that are abandoned or inefficiently used		Ratio of inefficiently used underground space to the total developed space
Consideration of UUS use (also adjacent) in building design	Expert rating	Share of buildings that were designed taking into account current or perspective UUS use
Proportion of energy efficient UUS premises		Ratio of total floor (non residential) area in the city to energy efficient floor area, share of underground of those
Proportion of buildings using UUS for energy sourcing or storage		Share of buildings using geothermal or shallow subsurface energy exchange to the total building stock
Innovations in infrastructure and UUS use	Expert rating	
UUS use density		Volume of all physical infrastructure assets located beneath city area, divided by respective area
UUS use by function		Share of different urban functions in UUS by occupied volume
Share of renewable energy by source, including underground		Share of geothermal energy in the total energy mix (may be limited to housing sector)
Average speed of passenger travel on underground public transport		Average speed as provided by e.g. public rail operator; can be compared to above ground speeds
Share of urban functions that are underground and not consuming urban land	Expert rating	Share of different urban functions in UUS as compared to above ground functions. Descriptor can be taken as annual turnover
Stratification of UUI by depth		Share of developed UUS according to depth occupied. Categories can be taken by e.g. 1 m

3.3. Data for urban underground space indicators

How can UUS indicators be measured? What can be the units? Is data available? Who collects and handles the data on UUS?

To discuss the above questions it is appropriate to give a brief historical overview of the subject. A number of scholarly articles mention *early use of underground spaces* giving location and date (e.g. Kaliampakos, 2015), but publications giving quantitative characteristics are comparatively rare.

An early publication introducing "volume of tunnelling" and "total [developed] subsurface volume of a city" appeared in 1976 in Underground Space journal (Jansson, 1976).Duffaut (1980) pioneered quantification of UUS use in relation to other city characteristics. In his article "Past and future use of underground space in France and Europe" Pierre Duffaut presented the total volume of developed UUS in Paris, and divided it by city area resulting in "average height of [developed] underground space under the entire Paris area" as 90 centimetres. Duffaut presented data on stratification of UUI by depth as well, breaking it into three categories (0–4 m depth; 4–10; and 10–50).

Anttikoski et al. (1989) presented detailed calculations of volumes of spaces comprising a subsurface public assembly hall in Helsinki, resulting in a total volume of 62,000 cubic metres.

Bobylev (2010b) made *a comprehensive underground assets inventory* of Alexanderplatz area in the centre of Berlin, Germany, using "volume of developed underground space" as the basic descriptor. In this study, share of urban functions "on land" and "underground" (respectively) were presented, as well as "UUI distribution by depth" and "volume of developed underground space per land area" were calculated.

He et al. (2012) attempted establishing *correlations* between "UUS density" and several urban indicators when analysing use of UUS in Shanghai, China. The strongest positive correlations were found between "UUS density" and "population density", "UUS density" and "real estate price"; weak correlation was found between "UUS density" and "Gross Domestic Product per capita". In this study, the indicator "real estate price" was represented by a proxy descriptor "annual turnover" of real estate market in respective districts of Shanghai adjusted to area size.

There are some *statistical studies* on UUS use as well. ITA (2012) indicates that there are 48 underground public rail transport systems in the world to date, while Broch (2016)estimates this figure as more than 100. Cui et al. (2013) presents analysis of 19 underground pedestrian systems in cities, citing the total number of those in the world as 51. Unfortunately this type of data hardly can be used in urban indicators research, since there is no basis for comparative analysis between cities or connection to any other urban characteristics which can be uniformly established. This highlights the challenges described in the previous sections regarding industry specific data use in urban indicators research.

A new initiative on the *Underground Atlas Project* (Kaliampakos et al., 2016) attempts to use crowdsourcing to create an online database of underground structures world-wide, which would help to expand knowledge *on geography of the structures* and eventually have some input of urban indicators cross-city analysis.

Table 6 summarises available quantitative data on UUS use, presenting a basis for UUS statistical analysis and UUS use monitoring as an urban indicator. Each of the UUS descriptors presented in Table 6 has particularities regarding its meaning and data availability options, which are summarised below:

Developed UUS volume can be sourced from construction companies, civil defence offices, or owners of the structures. Data on linear infrastructure and utilities can be calculated by sourcing data on its length and type (diameter) from utility companies.Bobylev (2010b) sourced data from 12 companies to calculate the volume under a half square kilometre area. Data on *cultural heritage (archaeological objects)* usually relies on expert judgement based on historical documents. There is a difference between inner and outer volume of the structures, which preferably should be taken into account. Most volume calculations would have a degree of *approximation*related to the above complexities. *City authorities* do not possess information on volumes of underground structures, no is there any other agency where this information can be found in one place (at least there were no reports worldwide suggesting the opposite to date). *The shallow subsurface* can be screened for *buried objects*, which is a contemporary scholarly and technologically challenging task (Metje et al., 2007) and is usually done during site investigation for new construction.

UUS use density descriptor obviously has the above described particularities pertaining to "volume of underground structures". Data on the other component of this descriptor, land area, is usually well documented and easily available from local authorities of respective city and its districts. Attention should be given to the situations where there are significant areas distinct from a considered urban area, e.g. unused sites, water bodies, which have no developed UUS and population. Inclusion of green areas is a difficult question, since large urban forests obviously can significantly alter results of densities for urban indicators.

Developed UUS volume per person descriptor is perhaps less vulnerable to distortion than "density of UUS use". Numbers of urban population are well documented and available. All urban population is using UUI (at least utilities), no matter of its exact *location*, hence the above described problems with spatially distributed indicators are avoided.

Underground premises floor area is easily available descriptor for any *commercial*premises, and constitutes the core of the real estate knowledge sector. The problem for UUS indicators research is that "floor area" is in most cases not sufficient to describe underground structures. This descriptor also is not applicable to utilities.

Table 6. Enabling data for Urban Underground Space (UUS) urban indicators.

Urban Underground Space (UUS) indicator/descriptor	Units	Units symbol
Developed UUS volume	Cubic metres	m^3
UUS use density	Cubic metres per land area	m^3/m^2
Developed UUS volume per person	Cubic metres per person (respective [urban] area population)	m^3/person
Underground premises floor area	Square metres	m^2

4. URBAN UNDERGROUND SPACE INDICATORS IN URBAN SUSTAINABILITY RESEARCH

This section attempts to put the above described ideas on UUS urban indicators into practice. There was no special data sourcing commissioned for this study, the information presented have been sourced from scholarly publications.

Table 7 summarises the information that has been found for this study. Data on "Developed UUS volume" was available for 8 cities, and for 3 of them the data was available in two different years. Indicators "UUS use density" and "Developed UUS volume per person" were calculated using data in the table. By UUS use density we mean estimation of the volume of all physical infrastructure assets located beneath city area. UUS assets include building basements, utility infrastructure, possible abandoned structures and cultural artefacts. UUS density, as population density, is shown as average for a city.

Fig. 1 presents the correlation between urban population and developed underground space densities. This graph represents *the best fit amongst the others plotted* usingTable 7 data. Cites (areas) shown here are Beijing, Helsinki, Paris, Quebec, Shanghai, and Stockholm. Central areas of Berlin and Paris are excluded, as well as projections for Beijing and Uddevalla. Hence, UUS use densities are limited here to below 0.05 m^3/m^2. Usually *city centres would have much more significant volumes of UUS*. Duffaut (2007)found that

Table 7. City and Urban Underground Space (UUS) use indicators in selected cities.

City	Area, km^2	Population ('000)	Population density, persons/km^2	Developed UUS volume, m^3 ('000 000)	UUS use density (m^3/m^2), shown in cm	Developed UUS volume per person m^3/person
Beijing, China 2006	3497	11,455	3275	90.0	2.6	7.9
Beijing, China 2020	3497	14,296	4088	270.0	7.7	18.9
Berlin (AlexB), Germany 2010	0.57			0.736	128.0	
Berlin (AlexS), Germany 2010	0.14			0.377	275.0	
Helsinki, Finland 1998	492	1117	2270	6.0	1.2	5.4
Helsinki, Finland 2013	215	600	2790	10.0	4.6	16.6
Paris (centre), France 1980	90	2211	24,566	82.0	91.1	37.1
Paris, France 2007	2845	10,485	3685	94.0	3.3	9.0
Quebec, Canada 1990	550	456	829	1.6	0.3	3.5
Shanghai, China 2012	6340	16,575	2614	120.0	1.9	7.2
Stockholm, Sweden 2005	382	1285	3363	11.3	2.9	8.8
Uddevalla, Sweden 1975	17	31	1823	0.35	2.1	11.3

Data sources:
City area: (Demographia, 2013), except for Uddevalla (www. uddevalla.se); Paris 1980 (Duffaut, 1980), Quebec (Boivin, 1990), Berlin (Bobylev, 2010b).
City population: (United Nations, 2012a).
Population density: (Demographia, 2013) or calculated using this table data.
Developed UUS volume: Uddevalla 1975 – (Jansson, 1976); Quebec 1990 – (Boivin, 1990); Helsinki 1998 – (Ronka et al., 1998); Helsinki 2013 – (Vahaaho, 2013); Shanghai (He et al., 2012); Beijing 2006, 2020 – (Shi, 2006); Paris 1980 – (Duffaut, 1980); Paris 2007 – (Duffaut, 2007); Stockholm 2005 – (Bobylev, 2008), Berlin (Bobylev, 201
Notes:
The year in the first column refers to when the data on UUS volume was estimated.
"UUS use density" and "Developed UUS volume per person" has been calculated using this table data.

Table 7 data was used to plot 6 diagrams, which are presented in Fig. 1, Fig. 2, Fig. 3,Fig. 4, Fig. 5 and Fig. 8.

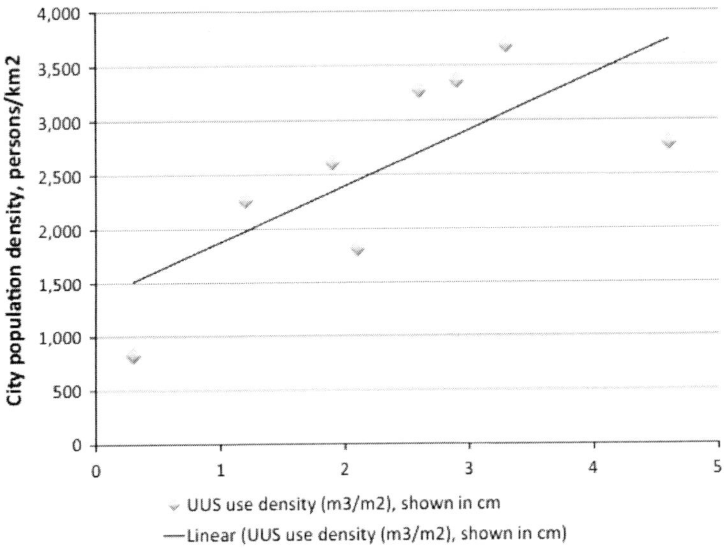

Fig. 1. Urban population and developed underground space densities.

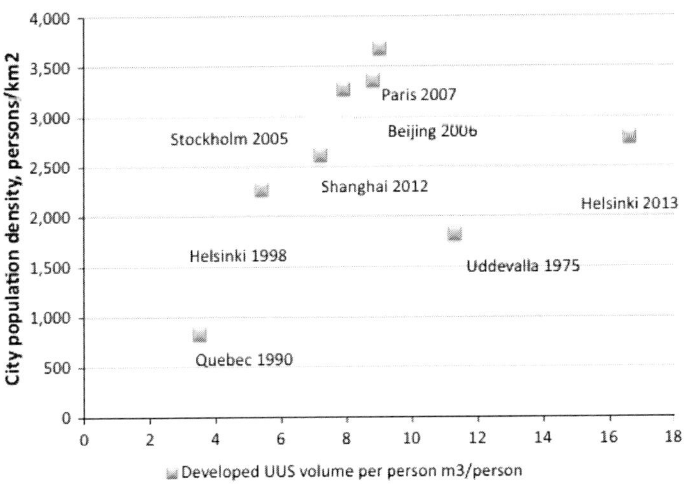

Fig. 2. Urban population density and developed underground space volume per person.

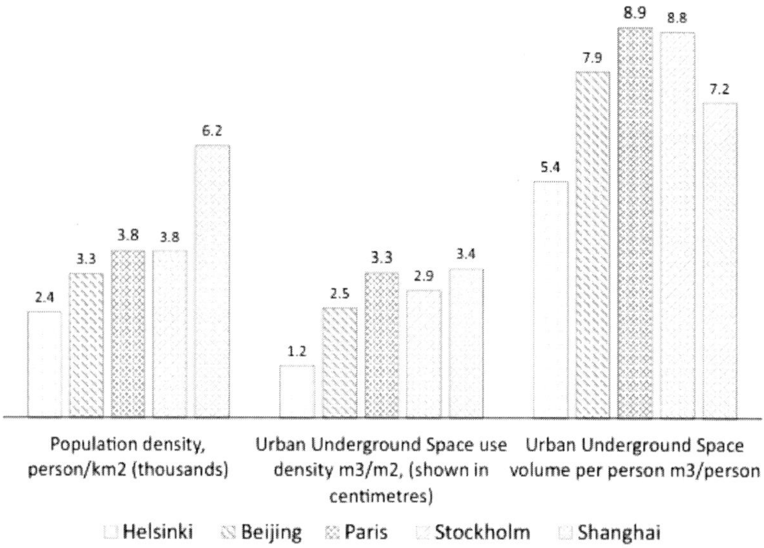

Fig. 3. Urban population density and two urban underground space use indicators in five cities.

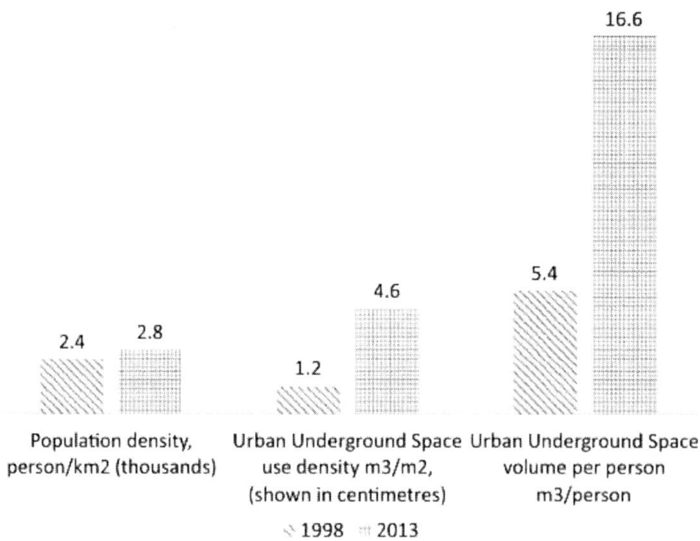

Fig. 4. Changes in population density and urban underground space use indicators in Helsinki.

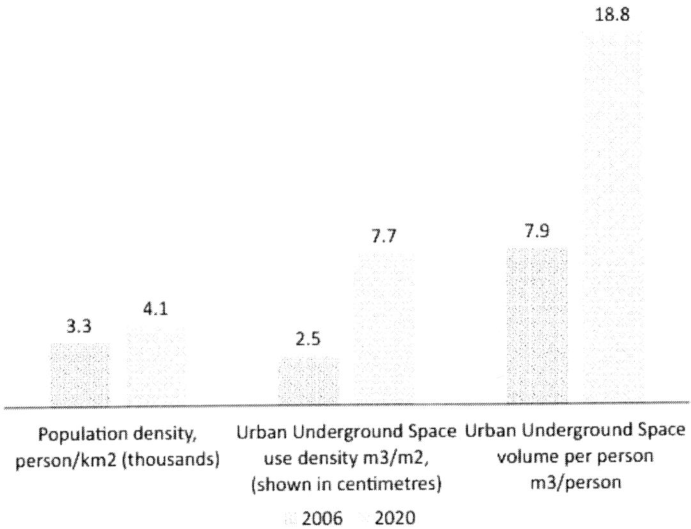

Fig. 5. Changes in population density and urban underground space use indicators in Beijing.

central area of Paris has on average about 91.1 cubic metre of underground structures' volume per square metre of land area (m^3/m^2), or an UUS virtual layer of just less than 1 m. Bobylev, 2010b, estimated this figure as *2.75* m^3/m^2 for the Alexanderplatz area in Berlin during a comprehensive case study. Average UUS densities for cities are much lower and are currently in the range of about *0.01–0.05* m^3/m^2.

Fig. 1 shows positive correlation between urban population and developed underground space densities for selected cities, which are developed cities that have significant financial resources. It can be interpreted that high population densities in developed cities are drivers for UUS use. Further explanations regarding this correlation can be found in the discussion section.

Fig. 2 presents correlation between urban population density and developed underground space volume per person, which *is less straightforward than the correlation shown in Fig. 1*. Still we can assume meaningful correlation, since the cities of Helsinki and

Uddevalla stand out by famously well developed underground space. Favourable ground conditions (bed rock) is a significant enabler for this development in both cities. Indicator UUS volume per person reflects UUS in a city not taking into account size of the urban area. This "population based" UUS indicator is a convenient way of reflecting on underground infrastructure which is used by all city dwellers, and might be located even outside the urban area (for example, sewerage treatment plants usually have sizable underground parts and often located outside city area).

Fig. 3 summarises urban population density and two UUS use indicators in five cities. This diagram presents no correlation, its purpose is to show the range of indicators values, and to show that they are in about one range for these diverse, but commonly large and important cities.

Fig. 4 and Fig. 5 make the best use of the available data and *compare UUS use in two cities*, Beijing and Helsinki. In both cities, data is available for two different years making a similar step of 15 years in Helsinki and 14 Beijing (the latter is projection to 2020 year). Interestingly, both cities show *rise in population densities*, which is not typical since wealthy developing cities are currently *sprawling* (assertion on the basis of the World Bank data (Angel et al., 2005)). One reading of this indicators comparison for Beijing and Helsinki is that both cities were able to maintain population densities growth by active use of UUS which had allowed further *densification* of the cities. Having in mind both cities and their respective country's development stage and context, we can further speculate that a need to accommodate growing population and provide public transportation is the driving force behind the Beijing UUS use growth; and *favourable ground conditions* for underground construction (bed rock) and available *wealth* facilitated UUS development in Helsinki. In these two cities we can see significant growth in UUS use, both in terms of land use density and its volume per person.

Fig. 6 and Fig. 7 give a more *local perspective on UUS use in a city centre*. The figures are sourced from *a comprehensive case study of UUS* use in two areas in the centre of Berlin around Alexanderplatz (Bobylev, 2010b). Berlin is, of course, a developed city with a rich history, the oldest documented underground structure in the case study areas was a sewer which dated back to 1880. Case study areas included the immediate vicinity of Alexanderplatz (small case study area 136 thousand square metres), and a larger area that covered several blocks of the city centre (big case study area 574 thousand square metres). As Fig. 6 shows, density of UUS use constitutes about 1–3 m^3/m^2, which can be considered as a typical representation of the current developed UUS densities in city centres of *developed or "mature" and wealthy cities.* Fig. 7 specifies distribution of underground infrastructure volume by depth, and highlights very intensive use of the first and the second metres depth of the shallow subsurface.

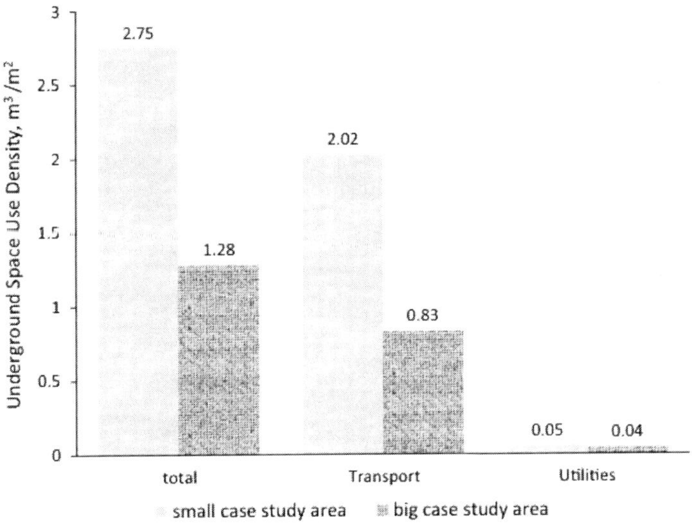

Fig. 6. Densities of the developed underground space in two case study areas. Underground infrastructure associated with Transport and Utility functions is presented as a part of all developed space. Case study of Alexanderplatz area, Berlin.

Source: Bobylev (2010b), edited.

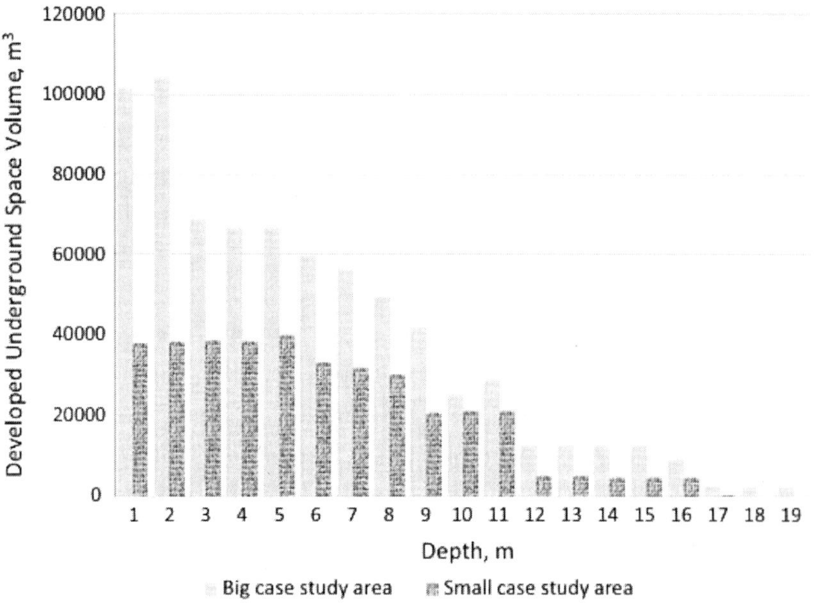

Fig. 7. Distribution of the developed underground space by depth. Case study of Alexanderplatz area, Berlin.

Fig. 8 is based on Table 8 and compares a city's prosperity index and two indicators of UUS use. *A City Prosperity Index* (UN Habitat, 2013a) is one of the similar integrated indexes which aim to capture all aspects of cites' life (e.g. by UNDP, UNECE, OECD, EUROSTAT, national governments). Urban prosperity is defined as a social construct that materialises in the realm of human actions (UN Habitat, 2013a), and a prosperous city is considered according to five main categories (productivity, infrastructure, quality of life, equity and social inclusion, environmental sustainability). According to the available statistics, there is weak correlation between city prosperity and UUS use, however we would argue that active UUS use is a sign of prosperity (which corresponds with UN Habitat "infrastructure" category), and, on the other hand, a city would need to achieve some level of prosperity to be financially able to significantly develop its UUS.

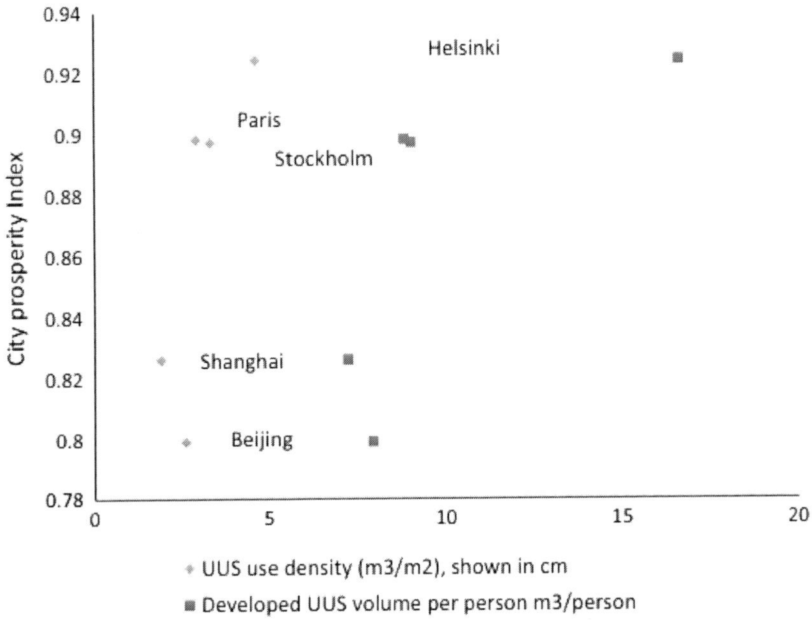

Fig. 8. City prosperity index and underground space use.

Presented data on urban UUS indicators is based on the scarce available statistics and is*indicative*, it might have a significant degree of inaccuracy. However it is important data that arguably estimates the direction and a scale of the urbanisation and UUS development processes accurately. It is worth mentioning the potential sources of inaccuracies in this research: (1) data on UUS represent credible estimations made by experts, there is no available comprehensive *"audit" of UUS assets*; (2) data on city areas and population can vary significantly, depending on *methodologies* of their estimation, here the challenges are whether city borders should be considered as those for administrative districts, and should empty lots, brownfields, parks, large water bodies be included. Reflecting further on data accuracy we can point out challenges to consider day and night population densities, and whether these should be based

Table 8. Urban Underground Space (UUS) use indicators and prosperity index in selected cities.

City	Area, km^2	Population ('000)	Population density, persons/km^2	Prosperity index	Developed UUS volume, m^3 ('000 000)	UUS use density (m^3/m^2), shown in cm	Developed UUS volume per person m^3/person
Beijing, China 2006	3497	11,455	3275	0.799	90.0	2.6	7.9
Helsinki, Finland 2013	215	600	2790	0.924	10.0	4.6	16.6
Paris, France 2007	2845	10,485	3685	0.897	94.0	3.3	9.0
Shanghai, China 2012	6340	16,575	2614	0.826	120.0	1.9	7.2
Stockholm, Sweden 2005	382	1285	3363	0.898	11.3	2.9	8.8

Data sources:
City area: (Demographia, 2013).
City population: (United Nations, 2012a).
Population density: (Demographia, 2013) or calculated using this table data.
Prosperity Index (with 5 dimensions): (UN-Habitat, 2013a).
Developed UUS volume: Helsinki 2013 – (Vahaaho, 2013); Shanghai (He et al., 2012); Beijing 2006 – (Shi, 2006); Paris 2007 – (Duffaut, 2007); Stockholm 2005 – (Bobylev, 2008).
Notes:
The year in the first column refers to when the data on UUS volume was estimated.
"UUS use density" and "Developed UUS volume per person" has been calculated using this table data.

on census or surveys of actual residence. A discussion on challenges for urban indicators data collecting can be found in e.g. World Bank, 2011. None the less, this data is *credible at the level of precision required* to analyse global trends and gasp the essence of the UUS indicators importance to urban development.

5. DISCUSSION

Urban indicators research is an important way to encourage and *monitor progress*towards sustainability. The data presented on UUS indicators, although still limited in numbers and geographically, is important *to mainstream UUS into urban sustainability research* and shows potential for *cross-disciplinary indicator analysis.* Table 9 suggests how UUS indicators are relevant to different levels or scales of their use.

Table 9. Levels/scale of urban indicators use and Urban Underground Space (UUS) relevance.

Indicator use scale	Urban Underground Space (UUS) relevance (arranged from top to bottom broadly corresponding to the Indicator categories from Global to Distinct Building)
Global	Urban sprawl
National	Land cover change
Regional	Urban energy efficiency (transport of people and goods)
Agglomeration	Rational land and underground planning
City	Rational use of urban space
City district	Efficiency of urban infrastructure
Municipality	Public open space
Block, street, square	Energy efficiency
Distinct building	

The analysed data show that global UUS use is a complex, *multi-dimensional* subject, which perhaps would not show any correlation or dependency upon other cities indicators that would be applicable to any city. This is true for the vast majority of urban indicators stemming from other disciplines as well, which is explained by the diverse nature of urban settlements.

Indeed, the *highest population densities* are observed in *poorly developed megacities*, often incorporating slum areas (e.g. Dhaka, Bangladesh has the highest density of 44,500 people per square kilometre (UN Habitat, 2013a)). These urban areas have hardly any underground infrastructure at all. However, in the developed cities, densification is required to make UUS use to be *economically feasible* (population densities in city centres, people per square kilometre: Helsinki – 16,500 (Vahaaho, 2013); Paris – 21,196 (INSEE, 2009)).

There is no direct correlation between city size in terms of area or population, city *wealth or prosperity*, and UUS indicators, although intensive UUS use is inherent to developed, wealthy countries and cities, or cities which are capitals and wealthy in comparison to others in the country (e.g. Moscow, Russia has elaborated metro and other UUS facilities).

An important factor that influences UUS development is, of course, *ground conditions*, upon which the price of underground excavations and construction risks largely depend. However unfavourable ground conditions (e.g. highly water saturated weak soils) cannot be nowadays seen as an obstacle for UUS development, since many e.g. European cities are built on weak soils and use UUS actively (e.g. Amsterdam).

UUS use in a city is a complex phenomenon and depends upon a large number of factors, including but not limited to ground conditions; climate; geographical location; urban area size in terms of area and population; variety of urban densities, including build stock, population, economic activity; policy and politics for using resources, including renewable energy and contemporary climate change agenda. Most notably, UUS use depends upon city development stage, i.e. transformation village–town–city, which is a complex notion in itself. City "maturity" is a necessary, but not sufficient condition for intensive UUS use.

6. CONCLUSION

6.1. Results

The paper has reviewed *urban indicator lists (systems)* and has identified the importance of UUS inclusion in the lists that give integrated assessment and monitor *urban sustainability, resilience, climate change adaptation and mitigation, as well as progress towards smart, liveable, and compact city*.

Urban indicators and the UUS themes are relevant to each other in two ways: (1) helping enhance *comprehensiveness and validity* of traditional urban indicator lists focused on sustainability, resilience, etc. (see Table 3 for details), and (2) helping better development of UUS, bringing *wider perspective on its resources use.*

Compatibility, comparability, uniformity, and sustained monitoring of urban indicators data (including UUS indicators) were found to be posing significant challenges to the research across geographies, and industry/economic sectors.

The best way to consider UUS in urban agendas is to include UUS relevant indicators at the *lowest hierarchy level* under the *established categories* in the lists. UUS can feature in several categories (see Table 4 for details).

Three key indicators (descriptors) are suggested to monitor UUS use:

- Developed UUS volume (m^3).
- UUS use density (m^3/m^2).
- Developed UUS volume per person (m^3/person).

Additionally data on *share of urban functions* in underground space and in above ground space can be collected, as well as stratification of UUS use by depth.

Current average UUS use densities in cities are up to about 0.05 (m^3/m^2) (which can be interpreted as a virtual depth of UUS use of 5 centimetres), and the developed UUS volume per person is up to about 10 m^3/person; while city central areas (central business districts) can have a virtual depth of the developed UUS of several metres (m^3/m^2).

Theoretically, the indicator "Developed UUS volume per person" should more accurately reflect on urban UUS use sine it avoids *conceptual problems* with defining the correct area for analysing *density;* however, the best fit correlation was found with the "UUS use density" indicator (Fig. 1).

6.2. Recommendations

- Enable co-ordination in developing methodological assistance, data collection, and monitoring of the *UUS urban indicators at an international level* (this role can be assumed by one of the *United Nations* agencies concerned with urban indicators). Similar co-ordination can be arranged at national levels as well.
- Create *inventory of UUS assets*, to make informed decisions on further development of UUS at the regional and city level. In fact this has been first highlighted by Sterling (1996) as a public policy recommendation stemming from a Colloquium on underground space utilisation in the US: "mandate the investigation and recording of an inventory of existing underground structures for future reference". This recommendation can be expanded to UUS resource use, including inventory of groundwater use and energy extraction/storage.
- Consider including UUS indicators in international *"green" building certification systems* like LEED and BREAM.

UUS is extremely relevant to the urban agendas, but without its *explicit inclusion into urban indicator lists*, UUS issues remain invisible to *policy makers.*

ACKNOWLEDGEMENT

The reported research has been funded by the European Community 7th Research Framework programme under People Marie Curie actions, Grant PIIF-GA-2010-273861.

REFERENCES

1. S. Angel, S.C. Sheppard, D.L. Civco **The Dynamics of Global Urban Expansion** World Bank, Washington, DC (2005)

2. U. Anttikoski, V. Grön, A. Halme, M. Juhola, M. Palmu, O. i Parkkamäki, T. Tuisku, O. Äyräväinen **Subsurface public assembly spaces in Finland** Tunn. Undergr. Space Technol., 4 (1) (1989), pp. 17–22 ISSN 0886-7798

3. N. Bobylev **Strategic environmental assessment of urban underground infrastructure development policies** Tunn. Undergr. Space Technol., 21 (3–4) (2006), pp. 469–479

4. N. Bobylev **Sustainability and vulnerability analysis of critical underground infrastructure** ,in: I. Linkov, R. Wenning, G. Kiker (Eds.), Managing Critical Infrastructure Risks, NATO Security through Science Series, Springer, Netherlands (2007), pp. 445–

5. N. Bobylev **Urbanization and environmental security: infrastructure development, environmental indicators, and sustainability** ,in: P.H. Liotta (Ed.), *et al.*, Environmental Change and Human Security, NATO Science Series: IV: Earth and Environmental Sciences, Springer Science + Business Media B.V. (2008), pp. 203–216 ISBN 978-1-4020-8550-5

6. N. Bobylev **Mainstreaming sustainable development into a city's master plan: a case of urban underground space use** Land Use Policy, 26 (4) (2009), pp. 1128–1137

7. Bobylev, N., 2009b. Urban underground infrastructure and climate change: opportunities and threats. In: 5th Urban

Research Symposium, Marseille, France, June 28–30, 2009, 17p

8. N. Bobylev **Ecosystem services in the context of human development** P.H. Liotta (Ed.), *et al.*, Achieving Environmental Security: Ecosystem Services and Human Welfare, IOS Press, Amsterdam (2010), pp. 183–205 ISBN 978-1-60750-578-5

9. N. Bobylev **Underground space use in the Alexanderplatz area, Berlin: research into the quantification of urban underground space use** Tunn. Undergr. Space Technol., 25 (5) (2010), pp. 495–507

10. N. Bobylev **Comparative analysis of environmental impacts of selected underground construction technologies using analytic network process** Automat. Constr., 20 (8) (2011), pp. 1030–1040

11. N. Bobylev **Urban physical infrastructure adaptation to climate change** ,in: J.B. Saulnier, M.D. Varella (Eds.), Global Change Energy Issues and Regulation Policies, Integrated Science & Technology Program 2, Springer Science+Business Media, Dordrecht (2013), pp. 77–102

12. Bobylev, N., 2014. Addressing urban disaster risks via a combination of responsible land use and critical infrastructure management. In: Presentation at the Geospatial World Forum. Land Information Systems for Smart Cities 8–9 May 2014, Geneva, Switzerland

13. Bobylev, N., Hunt, D.V.L., Jefferson, I., Rogers, C.D.F., 2013. Sustainable infrastructure for resilient urban environments. In: Zhou, Cai, Sterling (Eds.), Advances in Underground Space Development, Copyright 2013 by The Society for Rock Mechanics & Engineering Geology (Singapore). Published by Research Publishing, pp. 906–917. ISBN: 978-981-07-3757-3, http://dx.doi.org/10.3850/978-981-07-3757-3 (RP-107-P219).

14. D.J. Boivin **Underground space use and planning in the Québec City area** Tunn. Undergr. Space Technol., 5 (1–2) (1990), pp. 69–83

15. BRE Global, 2011. SD5065 Technical Guidance Manual: Version 1. BREEAM for Communities Assessor Manual: Development Planning Application Stage.

16. Broch, E., 2016. Planning and utilisation of rock caverns and tunnels in Norway. Tunn. Undergr. Space Technol., Special Issue on Urban Underground Space: A Growing Imperative. 55, 329–338.

17. Brundtland Commission, 1987. Our Common Future, Report of the World Commission on Environment and Development, World Commission on Environment and Development, 1987. Published as Annex to General Assembly Document A/42/427, Development and International Co-operation: Environment August 2, 1987.

18. J. Carmody, R. Sterling **Underground Space Design: A Guide to Subsurface Utilization and Design for People in Underground Spaces** Van Nostrand Reinhold, New York (1993) January 1993

19. G. Castanheira, L. Bragança **The evolution of the sustainability assessment tool SBToolPT: from buildings to the built environment** Sci. World J., 62 (4) (2014) 1 (2356–6140)

20. CEROI Project, 2010. Cities Environment Reports on the Internet (CEROI) Official Webpage <http://www.ceroi.net> (last visited: 22.3.2010).

21. CRISP Project, 1999. A European Thematic Network on Construction and City Related Sustainability Indicators. Final Report.

22. J. Cui, A. Allan, M. Taylor, D. Lin **Underground pedestrian systems development in cities: influencing factors and**

implications Tunn. Undergr. Space Technol., 35 (2013), pp. 152–160

23. G. Daly, A. González **Key indicators for territorial cohesion and spatial planning: the reform of the EU cohesion policy and the new role of spatial indicators** Borderl. J. Spat. Plan. Irel., 3 (2013), pp. 77–89

24. Demographia, 2013. World Urban Areas: 9th Annual Edition (2013.03) <http://www.demographia.com/>.

25. P. Duffaut **Past and future use of underground space in France and Europe** Undergr. Space, 5 (2) (1980), pp. 86–91

26. Duffaut, P., 2007. Urbanisme souterrain. Pano-rama historique et géographique. Techniques de l'Ingénieur C3061-2, Paris.

27. European Commission, 2003. European Common Indicators: Towards a Local Sustainability Profile. European Common Indicators Project, Final Project Report, Milano, Italy, May.

28. Eurostat, 2009. Sustainable Development in the European Union: 2009 Monitoring Report of the EU Sustainable Development Strategy. Eurostat Statistical Books.

29. Foster, N., 2011. Lord Foster Inaugural Lecture as Humanities Visiting Professor of Architecture at the University of Oxford

30. M.B. Gonzalo, M.D. Bovea, M.J. Ruá **Sustainability on the urban scale: proposal of a structure of indicators for the Spanish context** Environ. Impact Assess. Rev., 53 (2015) (2015), pp. 16–30

31. L. He, Y. Song, S. Dai, K. Durbak **Quantitative research on the capacity of urban underground space – the case of Shanghai, China** Tunn. Undergr. Space Technol., 32 (2012) (2012), pp. 168–179

32. Hunt, D.V.L., Makana, L.O., Jefferson, I., Rogers, C.D.F., 2016. Liveable cities and urban underground space. Tunn. Undergr. Space Technol., Special Issue on Urban Underground Space: A Growing Imperative. 55, 8–20.

33. INSEE, 2009. Institute National de la Statistique et des Études Économiques: INSEE (France's National Institute of Statistics and Economic Studies) online database

34. ITA, 2012. Report on Underground Solutions for Urban Problems. International Tunnelling and Underground Space Association. ISBN 978-2-9700776-5-7.

35. ITACUS, 2010. ITA Committee on Underground Space White Paper #2 "Planning the Use of Underground Space".

36. B. Jansson **Terraspace – a world to explore** Undergr. Space, 1 (1) (1976), pp. 9–18

37. Kaliampakos, D., 2015. New challenges in the use of underground space. SEE Tunnel: promoting tunneling in SEE Region. In: Proceedings of the ITA WTC 2015 Congress and 41st General Assembly, Dubrovnik, Croatia, May 22–28, 2015.

38. Kaliampakos, D., Benardos, A., Mavrikos, A., Panagiotopoulos, G., 2016. The Underground Atlas Project. Tunn. Undergr. Space Technol., Special Issue on Urban Underground Space: A Growing Imperative. 55, 229–235.

39. Lynch, A.J., Andreason, S., Eisenmen, T., Robinson, J., Stelf, K., Birch, E.L., 2011. Sustainable urban Development Indicators for the United States. Reports in Penn Penn Institute for Urban Research. White Paper Series on Sustainable Urban Development <

40. Makana, L.O., Jefferson, I., Hunt, D.V.L., Rogers, C.D.F., 2016. Assessment of the future resilience of urban sub-surface environments. Tunnel. Undergr. Space Technol., Special Issue

on Urban Underground Space: A Growing Imperative. 55, 21–31.

41. N. Metje, P.R. Atkins, M.J. Brennan, *et al.* **Mapping the underworld: state of the art review** Tunn. Undergr. Space Technol., 22 (5–6) (2007), pp. 568–586

42. F.L. Michael, Z.Z. Noor, M.J. Figueroa **Review of urban sustainability indicators assessment – case study between Asian countries** Habitat Int., 44 (2014) (2014), pp. 491–500

43. N. Moussiopoulos, C. Achillas, C. Vlachokostas, D. Spyridi, K. Nikolaou **Environmental, social and economic information management for the evaluation of sustainability in urban areas: a system of indicators for Thessaloniki, Greece** Cities, 27 (2010) (2010), pp. 377–384

44. OECD, 2004. Organization for Economic Co-operation and Development Key Environmental Indicators. OECD Environment Directorate, Paris, France.

45. OECD, 2006. Organization for Economic Co-operation and Development. Infrastructure to 2030: Telecom, Land Transport, Water and Electricity. ISBN 92-64-02398-4.

46. OECD, 2008. Organization for Economic Co-operation and Development. "Urbanisation", in OECD Environmental Outlook to 2030, OECD Publishing ISBN 978-92-64-04048-9.

47. Rogers, C.D.F., 2009. Substructures, underground space and sustainable urban environments. Engineering Geology for Tomorrow's Cities. In: Culshaw, M.G., Reeves, H.J., Jefferson, I., Spink, T.W. (Eds.), Geological Society, London, Engineering Geology Special Publications, No. 22, pp. 177–188.

48. K. Ronka, J. Ritola, K. Rauhala **Underground space in land-use planning** Tunn. Undergr. Space Technol., 13 (1) (1998), pp. 39–49

49. SCR, 2015. Sustainable Community Rating <www.places.vic.gov> (Last access June 2015).

50. L.Y. Shen, J.J. Ochoa, M.N. Shah, X. Zhang **The application of urban sustainability indicators – a comparison between various practices** Habitat Int., 35 (2011) (2011), pp. 17–29

51. Shi, X.D., 2006. Review and Vista of Beijing Underground Space Utilization. Planning and Construction of Beijing. Beijing Planning Review, Beijing, pp. 32–40.

52. Sterling, R., Nelson, P., 2013. City resiliency an underground space use. In: Zhou, Cai, Sterling (Eds.), Advances in Underground Space Development, Copyright 2013 by The Society for Rock Mechanics & Engineering Geology (Singapore). Published by Research Publishing. pp. 56–14. ISBN: 978-981-07-3756-6.

53. R. Sterling **Going under to stay on top, revisited: results of a colloquium on underground space utilization** Tunn. Undergr. Space Technol., 11 (3) (1996), pp. 263–270

54. R. Sterling **Underground technologies for liveable cities** Tunn. Undergr. Space Technol., 12 (4) (1997), pp. 7–8

55. R. Sterling, H. Admiraal, N. Bobylev, H. Parker, J.P. Godard, I. Vähäaho, C.D.F. Rogers, X. Shi, T. Hanamura **Sustainability issues for underground space in urban areas** Proc. ICE – Urban. Des. Plan., 165 (4) (2012), pp. 241–254

56. UN Habitat, 2006. The United Nations Human Settlements Programme. State of the World's Cities 2006/7, UN-HABITAT, Nairobi.

57. UN Habitat, 2013a. United Nations Human Settlements Programme. State of Cities Report. Routledge for and on Behalf of the United Nations Human Settlements Programme (UN-Habitat). ISBN13: 978-0-415-83888-7.

58. UN Habitat, 2013b. United Nations Human Settlements Programme. Urban indicators Guidelines: Monitoring the Habitat Agenda and the Millennium Development Goals

59. UNEP, 2012. United Nations Environment Programme. GEO5 Global Environment Outlook: Environment for the Future We Want. United Nations Environment Programme, Valletta, Malta.

60. United Nations, 1995. Agenda 21 <http://web.archive.org/web/20090420073232/> (last access, June 2015).

61. United Nations **Indicators of Sustainable Development: Guidelines and Methodologies** (third ed.)Department of Economic and Social Affairs, Commission on Sustainable Development, New York (2007)

62. United Nations, 2012a. World Population Prospects. Population Division of the United Nations Department of Economic and Social Affairs of the United Nations Secretariat <http://www.un.org/esa/desa>.

63. United Nations, 2012b. Secretary-General's Video Message to the World Tunnel Congress. Bangkok, Thailand, 21 May 2012 <http://www.un.org/sg/statements/?nid=6064> (Accessed on May 27).

64. United Nations Sustainable Development Office, 1998. Indicators of Sustainable Development: Guidelines and Methodologies.

65. Urban Audit, 2004. Urban Audit Methodological Handbook. Office for Official Publications of the European Communities.

66. US GBC, 2009. LEED for Neighborhood Development. The U.S. Green Building Council.

67. Vahaaho, I, 2013. 0-Land use: underground resources and master plan in Helsinki. In: Zhou, Cai, Sterling (Eds.), Advances in Underground Space Development, Copyright

2013 by The Society for Rock Mechanics & Engineering Geology (Singapore). Published by Research Publishing. pp. 56–14. ISBN: 978-981-07-3756-6.

68. W. Wende, W. Huelsmann, M. Marty, G. Penn-Bressel, N. Bobylev **Climate protection and compact urban structures in spatial planning and local construction plans in Germany** Land Use Policy, 27 (3) (2010), pp. 864–868

69. WHO, 1999. World Health Organization. Environmental Health Indicators: Framework and Methodologies. Protection of the Human Environment Occupational and Environmental Health Series, Geneva.

70. World Bank, 2011. The Changing Wealth of Nations: Measuring Sustainable Development in the New Millennium. World Bank, Washington, D.C..

71. World Bank, 2012. The World Bank Annual Report 2012: Volume 2. Responding with Knowledge and Experience. Washington, DC

72. World Bank, 2015. World Development Indicators 2015. World Bank, Washington, DC

3

CHAPTER

Design Methodology for a Quick and Low-Cost Wind Tunnel

Miguel A. González Hernández[1], Ana I. Moreno López[1], Artur A. Jarzabek[1], José M. Perales Perales[1], Yuliang Wu[2] and Sun Xiaoxiao[2]

[1] *Polytechnic University of Madrid, Spain*

[2] *Beijing Institute of Technology, China*

1. INTRODUCTION

Wind tunnels are devices that enable researchers to study the flow over objects of interest, the forces acting on them and their interaction with the flow, which is nowadays playing an increasingly important role due to noise pollution. Since the very first day, wind tunnels have been used to verify aerodynamic theories and facilitate the design of aircrafts and, for a very long time, this has remained their main application. Nowadays, the aerodynamic research has expanded into other fields such as automotive industry, architecture, environment, education, etc., making low speed wind tunnel tests more important. Although the usefulness of CFD methods has improved over time, thousands of hours of wind tunnel tests (WTT) are still essential for the development of a new aircraft, wind turbine or any other design that involves complex interactions with the flow. Consequently, due to the growing interest of other branches of industry and science in low speed aerodynamics, and due to the persistent incapability of achieving accurate solutions with numerical codes, low speed wind tunnels (LSWT) are essential and irreplaceable during research and design.

A crucial characteristic of wind tunnels is the flow quality inside the test chamber and the overall performances. Three main criteria that are commonly used to define them are: maximum achievable speed, flow uniformity and turbulence level. Therefore, the design aim of a wind tunnel, in general, is to get a controlled flow in the test chamber, achieving the necessary flow performance and quality parameters.

In case of the aeronautical LSWTs, the requirements of those parameters are extremely strict, often substantially increasing the cost of facilities. But low turbulence and high uniformity in the flow are only necessary when, for example, laminar boundary layers have to be investigated. Another example of their use is aircraft engines combustion testing; this in turns requires a costly system that would purify the air in the tunnel to maintain the same air quality. Another increasingly important part of aircraft design is their noise footprint and usually the only way to test this phenomenon is in a wind tunnel.

In the automotive applications, it is obvious that the aerodynamic drag of the car is of paramount importance. Nevertheless, with the currently high level of control of this parameter and also due to imposed speed limitations, most of the efforts are directed to reduce the aerodynamic noise. The ground effect simulation is also very important, resulting in very sophisticated facilities to allow testing of both the ground effect simulation and noise production in the test section.

In architecture, due to the fact that buildings are placed on the ground and are usually of relatively low height, they are well within the atmospheric boundary layer. Therefore, the simulation of the equivalent boundary layer, in terms of average speed and turbulence level, becomes a challenging problem.

The design of the wind tunnels depends mainly on their final purpose. Apart from vertical wind tunnels and others used for specific tests (e.g. pressurised or cryogenic wind tunnels), most of the LSWTs can be categorised into two basic groups: open and closed circuit. They can be further divided into open and closed test section type.

For most applications, mainly for medium and large size wind tunnels, the typical configuration is the closed circuit and closed test chamber. Although, due to the conservation of kinetic energy of the airflow, these wind tunnels achieve the highest economic operation efficiency, they prove more difficult to design resulting from their general complexity. Hence, we will pay more attention to them in this chapter.

Apart from some early built wind tunnels for educational purposes at the UPM, since 1995 a number of LSWTs have been designed following the methodology which will be presented here. It focuses on the reduction of construction and operation costs, for a given performance and quality requirements.

The design procedure was first used for a theoretical design of a LSWT for the Spanish Consejo Superior de Deportes, which was to have a test section of 3,0 x 2,5 x 10,0 m^3 with a maximum operating speed of 40 m/s. Based on this design, a 1:8 scale model was built at UPM. This scaled wind tunnel has been used for research and educational purposes.

The second time it was during the design of a LSWT for the Instituto Tecnológico y de Energías Renovables de Tenerife (ITER). That wind tunnel is in use since February 2001, operating in two configurations: medium flow quality at maximum operating speed of 57 m/s, and high flow quality at maximum operating speed of 48 m/s. For more information visit www.iter.es.

Another example of this design procedure is a LSWT for the Universidad Tecnológica de Perú, which is now routinely used for teaching purposes. This wind tunnel is now in operation for about one and a half year.

At the moment the same procedure is being utilised to design a LSWT for the Beijing Institute of Technology (BIT). This wind tunnel will be used for educational and research purposes. It will have a high quality flow, up to 50 m/s, in a test section of 1,4 x 1,0 x 2,0 m^3. It will be used for typical aerodynamic tests and airfoil cascade tests (utilising the first corner of the wind tunnel circuit).

The design method to be presented in this chapter is based upon classical internal ducts design and analysis method, e.g. *Memento des pertes de charge: Coefficients de pertes de charge singulières et de pertes de charge par frottement,* I.E. Idel'cik [Eyrolles, 1986]. It also includes design assisting software such as a macro-aided Excel spreadsheet with all the complete formulation and dimensioning schemes for automatic recalculation. At the moment the best example of use of the method is the BIT-LSWT, mentioned above, as it has been defined using the latest and most reliable generation of wind tunnel design methodology.

2. MAIN DESIGN CRITERIA

The general layout of the proposed wind tunnel is shown in Figure 1. The airflow circulates in the direction indicated in the test chamber (counter clockwise in the figure). Upstream of the test chamber we find the other two main components of the wind tunnel: the contraction zone and the settling chamber. The other crucial component is of course the power plant. The remainder of the components just serve the purpose of closing the circuit while minimising the pressure loss. Nevertheless, diffuser 1 and corner 1 also have an important influence on the flow quality and they are responsible for more than 50% of the total pressure loss.

The design criteria are strongly linked with the specifications and requirements and those must be in accordance with the wind tunnel applications. The building and operation costs of a wind tunnel are highly related to the specifications and these are just a consequence of the expected applications.

In the case of the so called Industrial Aerodynamics or educational applications, the requirements related to flow quality may be relaxed, but for research and aeronautical applications the flow quality becomes very important, resulting in more expensive construction and higher operational costs.

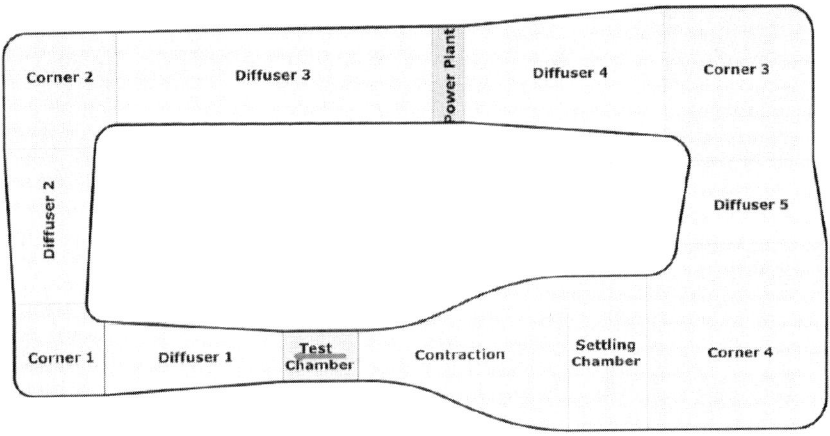

FIGURE 1. General layout of a closed circuit low speed wind tunnel. Figure labels indicate the part name, according to standards.

The main specifications for a wind tunnel are the dimensions of the test section and the desired maximum operating speed. Together with this the flow quality, in terms of turbulence level and flow uniformity, must be specified in accordance with the applications. At this point it should also be defined whether all the components of the wind tunnel are going to be placed on the floor in a horizontal arrangement or in a vertical one, with only half of the circuit on the floor and the other half on top of it.

Flow quality, which is one of the main characteristics, is a result of the whole final design, and can only be verified during calibration tests. However, according to previous empirical knowledge, some rules can be followed to select adequate values of the variables that affect the associated quality parameters. The recommended values will be discussed in the sections corresponding to the Contraction,

Settling Chamber, Diffusor 1 and Corner 1, which are the wind tunnel parts that have the greatest impact on the flow quality.

Once these specifications are given, it is very important to obtain on one side the overall wind tunnel dimensions to check their compatibility with the available room, and on the other side a preliminary estimation of the overall cost. The cost is mainly associated to the external shape of the wind tunnel and the power plant requirements.

For the benefit of new wind tunnel designers, a tool has been devised and implemented in an Excel spreadsheet (visit web page http://www.aero.upm.es/LSLCWT). Using this tool the designer will immediately get information about each part of the wind tunnel, the overall dimensions, the global and individual pressure loss coefficients, and the required power. This will be done according to the recommended input parameters and specification based on the intended use of the wind tunnel.

3. WIND TUNNEL COMPONENTS DEFINITION

In the following sections the design of each part will be thoroughly discussed and analysed in detail to get the best design addressing the general and particular requirements. Before dealing with each component, some general comments are given for the most important parts. In the case of the contraction zone, its design is crucial for achieving the required flow quality in the test section. In this sense, its contraction ratio, length and contour definition determine the level of uniformity in the velocity profile, as well as the necessary turbulence attenuation. It is crucial to avoid flow separation close to the walls of the contraction zone. At the stage of design, the most adequate method to verify that design meets those criteria is computational fluid dynamics (CFD).

Other important parts of the wind tunnel design worth mentioning here are the corners which incorporate turning vanes. Their aim is to reduce pressure loss and, in the case of the corner 1, possibly improve flow quality in the test section. The parameters to be considered in their design are the spacing between vanes (whether the space ought to be constant or not) and the possibility of expanding the flow (increasing the cross-section).

To complete the design process, the measurement equipment needs to be defined together with the complimentary calibration tests. Special attention needs to be devoted to the specification and selection of the balance for forces measurement, a device that is used to measure aerodynamic forces and moments on the model subjected to airflow in the test section. Since the drag force on test subjects can be very small and significant noise may be coming from the vibration of the tunnel components, such as the model stand, the true drag value may become obscured. The choice of an appropriate force balance is therefore crucial in obtaining reliable and accurate measurements.

The selection depends mainly on the nature of the tests. Wind tunnel balances can be categorized into internal and external ones. The former offers mobility since it is usually only temporarily mounted to the test section and may be used in different test sections. However, the latter has more potential in terms of data accuracy and reliability since it is tailored to a specific wind tunnel and its test section. Due to this reason, external force balances should be studied in greater depth.

3.1. TEST CHAMBER

The test chamber size must be defined according to the wind tunnel main specifications, which also include the operating speed and desired flow quality. Test chamber size and operating speed determine the maximum size of the models and the maximum achievable Reynolds number.

The cross-section shape depends on the applications. In the case of civil or industrial applications, in most of the cases, a square cross-section is recommended. In this case, the test specimens are usually bluff bodies and their equivalent frontal area should not be higher than 10% of the test chamber cross-sectional area in order to avoid the need of making non-linear blockage corrections. Accurate methods for blockage corrections are presented in Maskell (1963).

Nevertheless, a rectangular shape is also recommended for aeronautical applications. In the case of three-dimensional tests, a typical width to height ratio is 4:3; however, for two-dimensional tests a 2:5 ratio is advised in order for the boundary layer thickness in the test section to be much smaller than the model span.

Taking into account that it is sometimes necessary to place additional equipment, e.g. measuring instruments, supports, etc., inside the test chamber, it is convenient to maintain the operation pressure inside it equal to the local environment pressure. To fulfil this condition, it is recommended to have a small opening, approximately 1,0% of the total length of the test chamber, at the entrance of the diffuser 1.

From the point of view of the pressure loss calculation, the test chamber will be considered as a constant section duct with standard finishing surfaces. Nevertheless, in some cases, the test chamber may have slightly divergent walls, in order to compensate for the boundary layer growth. This modification may avoid the need for tail flotation correction for aircraft model tests, although it would be strictly valid only for the design Reynolds number.

Figure 2 shows a design of a typical constant section test chamber. With the typical dimensions and velocities inside a wind tunnel, the flow in the test section, including the boundary layer, will be turbulent, because it is continuous along the whole wind tunnel. According to Idel´Cik (1969), the pressure loss coefficient, related to the dynamic pressure in the test section, which is considered as the reference dynamic pressure for all the calculations, is given by the expression:

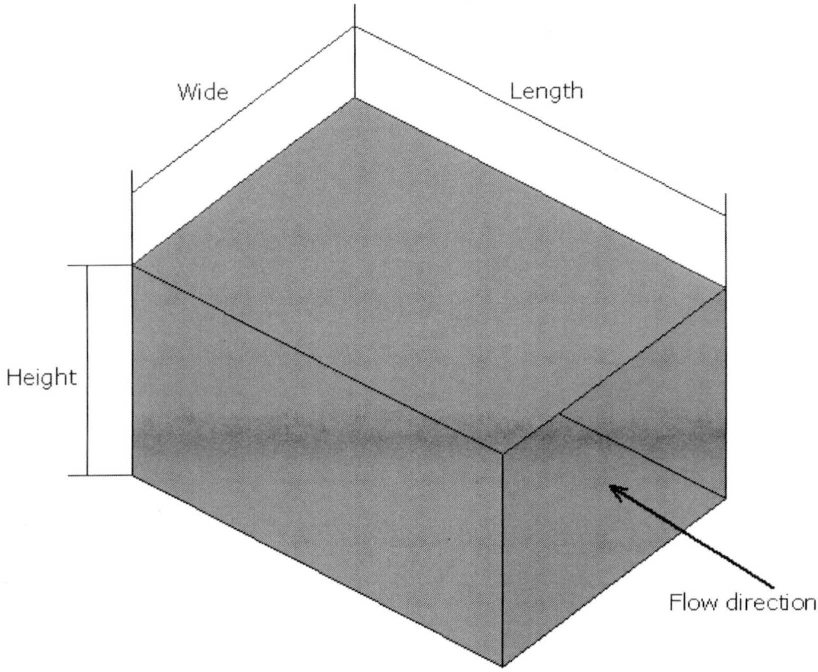

FIGURE 2. Layout of a constant section wind tunnel test chamber.

$$\zeta = \lambda \cdot L \;/ D_H,$$

where L is the length of the test chamber, D_H the hydraulic diameter and λ a coefficient given by the expression:

$$\lambda = 1/(1{,}8 \cdot \log Re - 1{,}64)^2,$$

where Re is the Reynolds number based on the hydraulic diameter.

3.2. CONTRACTION

The contraction or "nozzle" is the most critical part in the design of a wind tunnel; it has the highest impact on the test chamber flow quality. Its aim is to accelerate the flow from the settling chamber to the test chamber, further reducing flow turbulence and non-uniformities in the test chamber. The flow acceleration and non-uniformity attenuations mainly depend on the so-called contraction ratio, *N*, between the entrance and exit section areas. Figure 3 shows a typical wind tunnel contraction.

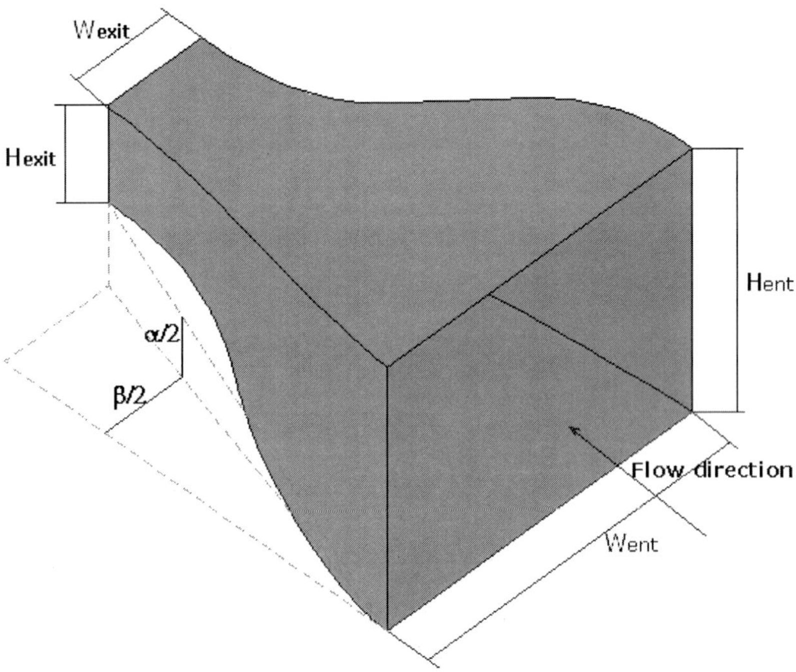

FIGURE 3. General layout of a three-dimensional wind tunnel contraction.

Although, due to the flow quality improvement, the contraction ratio, *N*, should be as large as possible, this parameter strongly influences the overall wind tunnel dimensions. Therefore, depending on the expected applications, a compromise for this parameter should be reached.

Quoting P. Bradshaw and R. Metha (1979), "The effect of a contraction on unsteady velocity variations and turbulence is more complicated: the reduction of x-component (axial) fluctuations is greater than that of transverse fluctuations. A simple analysis due to Prandtl predicts that the ratio of root-mean-square (rms) axial velocity fluctuation to mean velocity will be reduced by a factor $1/N^2$, as for mean-velocity variations, while the ratio of lateral rms fluctuations to mean velocity is reduced only by a factor of N: that is, the lateral fluctuations (in m/s, say) increase through the contraction, because of the stretching and spin-up of elementary longitudinal vortex lines. Batchelor, *The Theory of Homogeneous Turbulence*, Cambridge (1953), gives a more refined analysis, but Prandtl's results are good enough for tunnel design. The implication is that tunnel free-stream turbulence is far from isotropic. The axial-component fluctuation is easiest to measure, e.g. with a hot-wire anemometer, and is the "free-stream turbulence" value usually quoted. However, it is smaller than the others, even if it does contain a contribution from low-frequency unsteadiness of the tunnel flow as well as true turbulence."

In the case of wind tunnels for civil or industrial applications, a contractions ratio between 4,0 and 6,0 may be sufficient. With a good design of the shape, the flow turbulence and non-uniformities levels can reach the order of 2,0%, which is acceptable for many applications. Nevertheless, with one screen placed in the settling chamber those levels can be reduced up to 0,5%, which is a very reasonable value even for some aeronautical purposes.

For more demanding aeronautical, when the flow quality must be better than 0,1% in non-uniformities of the average speed and longitudinal turbulence level, and better than 0,3% in vertical and lateral turbulence level, a contraction ratio between 8,0 and 9,0 is more desirable. This ratio also allows installing 2 or 3 screens in the settling chamber to ensure the target flow quality without high pressure losses through them.

The shape of the contraction is the second characteristic to be defined. Taking into account that the contraction is rather smooth, one may think that a one-dimensional approach to the flow analysis would be adequate to determine the pressure gradient along it.

Although this is right for the average values, the pressure distribution on the contraction walls has some regions with adverse pressure gradient, which may produce local boundary layer separation. When it happens, the turbulence level increases drastically, resulting in poor flow quality in the test chamber.

According to P. Bradshaw and R. Metha (1979), “The old-style contraction shape with a small radius of curvature at the wide end and a large radius at the narrow end to provide a gentle entry to the test section is not the optimum. There is a danger of boundary-layer separation at the wide end, or perturbation of the flow through the last screen. Good practice is to make the ratio of the radius of curvature to the flow width about the same at each end. However, a too large radius of curvature at the upstream end leads to slow acceleration and therefore increased rate of growth of boundary-layer thickness, so the boundary layer - if laminar as it should be in a small tunnel - may suffer from Taylor-Goertler "centrifugal" instability when the radius of curvature decreases”.

According to our experience, when both of the contraction semi-angles, $\alpha/2$ and $\beta/2$ (see Figure 3), take the values in the order of 12º, the contraction has a reasonable length and a good fluid dynamic behaviour. With regard to the contour shape, following the recommendations of P. Bradshaw and R. Metha (1979), two segments of third degree polynomial curves are recommended.

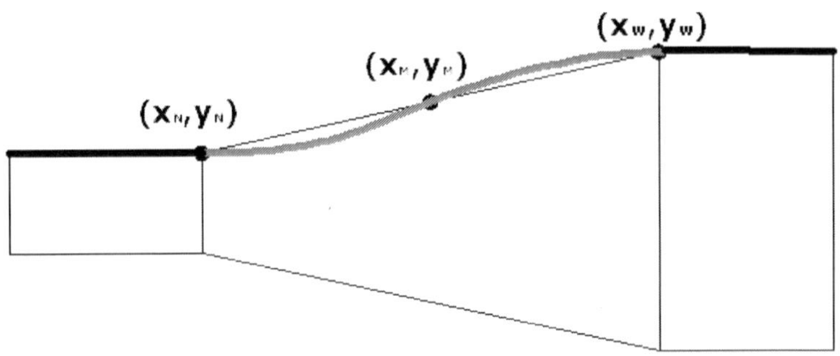

FIGURE 4. Fitting polynomials for contraction shape.

As indicated in Figure 4, the conditions required to define the polynomial starting at the wide end are: the coordinates (x_W, y_W), the horizontal tangential condition in that point, the point where the contour line crosses the connection strait line, usually in the 50% of such line, and the tangency with the line coming from the narrow end. For the line starting at the narrow end the initial point is (x_N, y_N), with the same horizontal tangential condition in this point, and the connection to the wide end line. Consequently, the polynomials are:

$$y = a_W + b_W \cdot x + c_W \cdot x^2 + d_W \cdot x^3,$$

$$y = a_N + b_N \cdot x + c_N \cdot x^2 + d_N \cdot x^3.$$

Imposing the condition that the connection point is in the 50%, the coordinates of that point are $[x_M, y_M] = [(x_W + x_N)/2, (y_W + y_N)/2)]$. Introducing the conditions in both polynomial equations, the two families of coefficients can be found.

According to Idel´Cik (1969), the pressure loss coefficient related to the dynamic pressure in the narrow section, is given by the expression:

$$\zeta = \left\{\frac{\lambda}{\left[16 \cdot \sin\left(\frac{\alpha}{2}\right)\right]}\right\}\left(1 - \frac{1}{N^2}\right) + \left\{\frac{\lambda}{\left[16 \cdot \sin\left(\frac{\beta}{2}\right)\right]}\right\}\left(1 - \frac{1}{N^2}\right),$$

where λ is defined as:

$$\lambda = 1 / (1{,}8 \log Re - 1{,}64)^2.$$

The Reynolds number is based on the hydraulic diameter of the narrow section.

3.3. SETTLING CHAMBER

Once the flow exits the fourth corner (see Figure 1), the uniformization process starts in the settling chamber. In the case of low-quality flow requirements, it is a simple constant section duct, which connects the exit of the corner 4 with the entrance of the contraction.

Nevertheless, when a high quality flow is required, some devices can be installed to increase the flow uniformity and to reduce the turbulence level at the entrance of the contraction (see Figure 5). The most commonly used devices are screens and honeycombs. Both devices achieve this goal by producing a relatively high total pressure loss; however, keeping in mind that the local dynamic pressure equals to $1/N^2$ of the reference dynamic pressure, such pressure loss will only be a small part of the overall one, assuming that N is large enough.

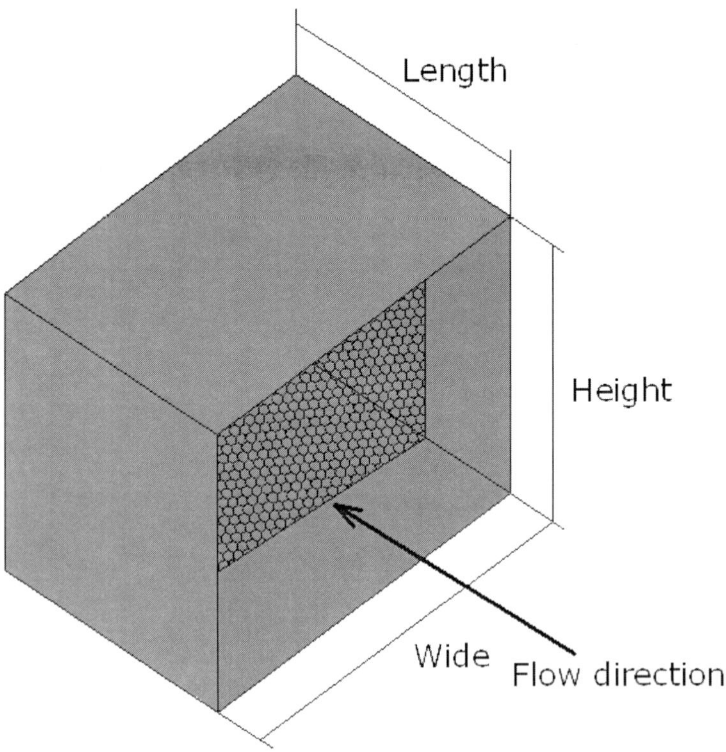

FIGURE 5. General layout of a settling chamber with a honeycomb layer.

Honeycomb is very efficient at reducing the lateral turbulence, as the flow pass through long and narrow pipes. Nevertheless, it introduces axial turbulence of the size equal to its diameter, which restrains the thickness of the honeycomb. The length must be at least 6 times bigger than the diameter. The pressure loss coefficient, with respect to the local dynamic pressure, is about 0,50 for a 3 mm diameter and 30 mm length honeycomb at typical settling chamber velocities and corresponding Reynolds numbers.

Although screens do not significantly influence the lateral turbulence, they are very efficient at reducing the longitudinal turbulence. In this case, the problem is that in the contraction chamber the lateral turbulence is less attenuated than the longitudinal one. As mentioned above, one screen can reduce very drastically the longitudinal turbulence level; however, using a series of 2 or 3 screens can attenuate turbulence level in two directions up to the value of 0,15%. The pressure loss coefficient, with respect to the local dynamic pressure, of an 80%-porous screen made of 0,5 mm diameter wires is about 0,40.

If a better flow quality is desired, a combination of honeycomb and screens is the most recommended solution. This configuration requires the honeycomb to be located upstream of 1 or 2 screens. In this case, the pressure loss coefficient, with respect to the local dynamic pressure, is going to be about 1,5. If the contraction ratio is 9, the impact on the total pressure loss coefficient would be about 0,02, which may represents a 10% of the total pressure loss coefficient. This implies a reduction of 5% in the maximum operating speed, for a given installed power.

The values of the pressure loss coefficients given in this section are only approximated and serve as a guideline for quick design decisions. More careful calculations are recommended for the final performance analysis following Idel´Cik's (1969) methods.

3.4. DIFFUSERS

The main function of diffusers is to recover static pressure in order to increase the wind tunnel efficiency and, of course, to close the circuit. For that reason, and some other discussed later, it is important to maintain the flow attachment for pressure recovery efficiency. Figure 6 shows the layout of a rectangular section diffuser.

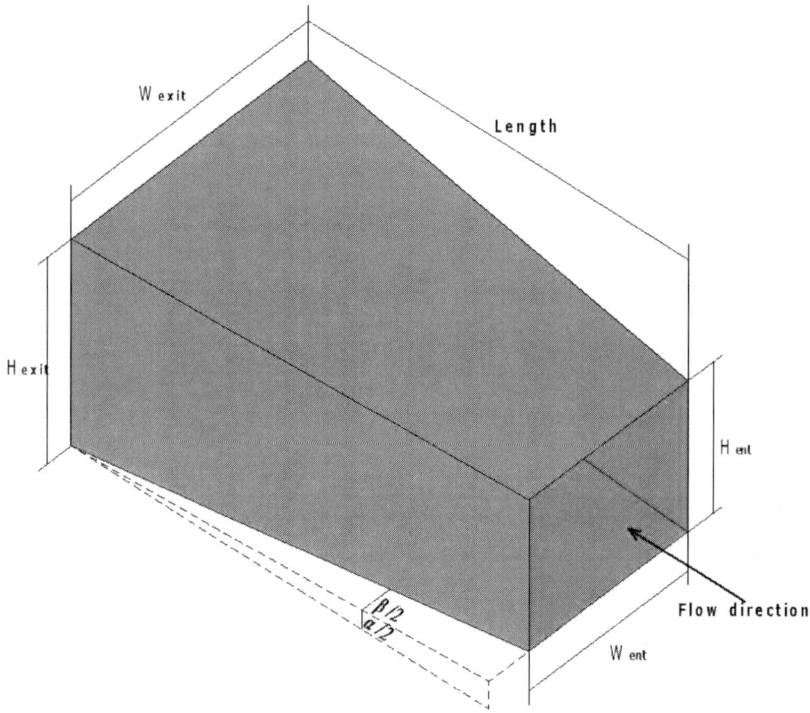

FIGURE 6. Rectangular section diffuser.

Diffuser 1 pays an important role in the test chamber flow quality. In case of flow detachment, the pressure pulsation is transmitted upstream into the test chamber, resulting in pressure and velocity non-uniformities. In addition, diffuser 1 acts as a buffer in the transmission of the pressure disturbances generated in the corner 1.

It has been proved that in order to avoid flow detachment, the maximum semi-opening angle in the diffuser has to be smaller than 3,5°. On the other hand, it is important to reduce as much as possible the dynamic pressure at the entrance of the corner 1, in order to minimise the possible pressure loss. Consequently, it is strongly recommended not to exceed the semi-opening angle limit and to design the diffuser to be as long as possible.

Diffuser 2 is a transitional duct, where the dynamic pressure is still rather large. Subsequently, the design criterion imposing a maximum value of the semi-opening angle must also be applied. The length of this diffuser cannot be chosen freely, because later it becomes restrained by the geometry of corners 3 and 4 and diffuser 5.

Diffuser 3 guides the flow to the power plant which is strongly affected by flow separation. In order to avoid it, the criterion imposing a maximum value of the semi-opening angle is maintained here as well. The cross-sectional shape may change along this diffuser because it must connect the exit of corner 2, whose shape usually resembles that of the test chamber, with the entrance of the power plant, whose shape will be discussed later.

The same can be said about diffuser 4 because pressure oscillations travel upstream and therefore may affect the power plant. Analogically to the previous case, it provides a connection between the exit of the power plant section and the corner 3, which has a cross-section shape resembling the one of the test chamber.

Diffuser 5 connects the corners 3 and 4. It is going to be very short, due to a low value of the dynamic pressure, which will allow reducing the overall wind tunnel size. This will happen mainly when the contraction ratio is high and the diffusion angle may be higher than 3,5°. It can also be used to start the adaptation between the cross-section shapes of the tests section and the power plant.

An accurate calculation of the pressure loss coefficient can be done with Idel´Cik´s (1969) method. A simplified procedure, derived from the method mentioned above, is presented here to facilitate a quick estimation of such coefficient.

The pressure loss coefficient, with respect to the dynamic pressure in the narrow side of the diffuser, is given by:

$$\zeta = 4{,}0 \cdot \tan \alpha / 2 \cdot \sqrt[4]{\tan {}^{\alpha}\!/_{2}} \cdot \left(1 - \frac{F_0}{F_1}\right)^2 + \zeta_f .$$

α being the average opening angle, F_0 the area of the narrow section, F_1 the area of the wide section and where ζfζf is defined as:

$$\zeta_f = \frac{0{,}02}{8 \cdot \sin \alpha / 2}\left[1 - \left(\frac{F_0}{F_1}\right)^2\right].$$

3.5. CORNERS

Closed circuit wind tunnels require having four corners, which are responsible for more than 50% of the total pressure loss. The most critical contribution comes from the corner 1 because it introduces about 34% of the total pressure loss. To reduce the pressure loss and to improve the flow quality at the exit, corner vanes must be added. Figure 7 shows a typical wind tunnel corner, including the geometrical parameters and the positioning of corner vanes.

The width and the height at the entrance, W_{ent} and H_{ent} respectively, are given by the previous diffuser dimensions. The height at the exit, H_{exit}, should be the same as at the entrance, but the width at the exit, W_{exit}, can be increased, giving the corner an expansion ratio, W_{exit}/W_{ent}. This parameter can have positive effects on the pressure loss coefficient of values up to approximately 1,1. However, it must be designed considering specific geometrical considerations, which will be discussed, in greater details in the general arrangement.

The corner radius is another design parameter and it is normally proportional to the width at the corner entrance. The radius will be identical for the corner vanes. Although increasing the corner radius reduces the pressure loss due to the pressure distribution on corner vanes, it increases both the losses due to friction and the overall wind tunnel dimensions. According to previous experience, it is recommended to use 0,25 W_{ent} as the value of the radius for corners 1 and 2, and 0,20 W_{ent} for the other two corners.

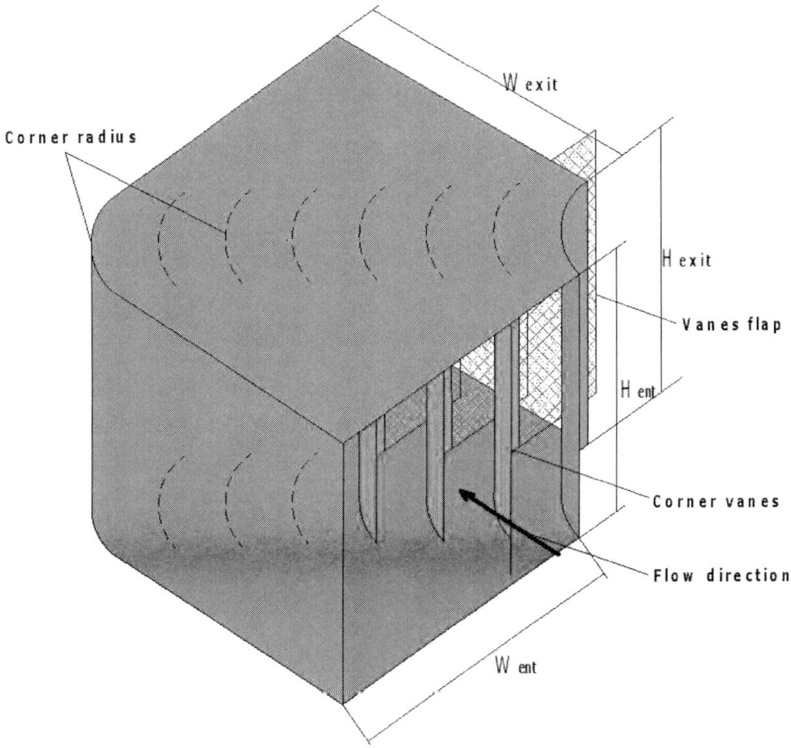

FIGURE 7. Scheme of a wind tunnel corner, including vanes, flaps and nomenclature.

The corner vanes spacing is another important design parameter. When the number of vanes increases, the loss due to pressure decreases, but the friction increases. Equal spacing is easier to define and sufficient for all corners apart from corner 1. In this case, in order to minimise pressure loss, the spacing should be gradually increased from the inner vanes to the outer ones.

The vanes can be defined as simple curved plates, but they can also be designed as cascade airfoils, which would lead to further pressure loss reduction. In the case of low speed wind tunnels the curved plates give reasonably good results. However, corner 1 may require to further stabilise the flow and reduce the pressure loss. Flap extensions with a length equal to the vane chord, as shown in Figure 7, is a strongly recommended solution to this problem.

Other parameters, such as the arc length of the vanes or their orientation, are beyond the scope of this chapter. For more thorough approach the reader should refer to Idel´Cik (1969), Chapter 6. As mentioned above, the pressure loss reduction in the corners is very important. Therefore, an optimum design of these elements, at least in the case of corner 1 and 2, has a significant impact on the wind tunnel performance.

In order to allow a preliminary estimation of the pressure loss in the corners we will follow the method presented in Diagram 6.33 from Idel´Cik (1969) mentioned above. In this approach, we take an average number of vanes, $n = 1{,}4^{*}S/t_1$, S being the diagonal dimension of the corner, where t_1 is the chord of the vane. The pressure loss coefficient is given by the expression:

$$\zeta = \zeta_M + 0{,}02 + 0{,}031^{*}\frac{r}{W_{ent}}.$$

ζ_M depends on r/W_{ent}, and its values are 0,20 and 0,17 for r/W_{ent} equal to 0,20 and 0,25, respectively. As a result, the corresponding values of ζ are 0,226 and 0,198 respectively, always with respect to the dynamic pressure at the entrance. This proves the validity of the recommendations given before with regard to the value of the curvature radius and the length of diffusor 1.

3.6. POWER PLANT

The main aim of the power plant is to maintain the flow running inside the wind tunnel at a constant speed, compensating for all the losses and dissipation. The parameters that specify it are the pressure increment, Δp, the volumetric flow, Q, and the power, P. Once the test chamber cross-section surface, S_{TC}, and the desired operating speed, V, are fixed, and the total pressure loss coefficient, ζ, has been calculated, all those parameters can be calculated using:

$$\Delta p = \frac{1}{2}\rho \cdot V^2 \cdot \zeta$$

$$Q = V \cdot S_{TC}$$

$$P = \Delta p \cdot \frac{Q}{\eta},$$

where ρ is the operating air density and η the fan efficiency, accounting for both aerodynamic and electric motor efficiencies.

In order to reduce the cost of this part by roughly one order of magnitude, we propose to use a multi-ventilator matrix, as presented in Figure 8, instead of a more standard single ventilator power plant configuration. The arrangement of this matrix will be discussed later.

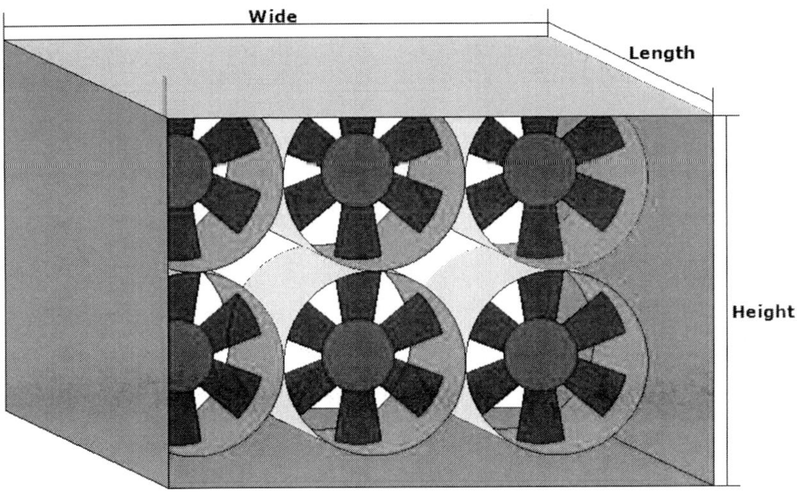

FIGURE 8. Layout of a multi-fan power plant.

According to our experience, for a closed circuit wind tunnel eventually including settling chamber screens or/and a honeycomb, the total pressure loss coefficient is in the range of 0,16 to 0,24. Consequently, in the case of 1,0 m^2 test section area and 80 m/s maximum operating speed, assuming an average value of ζ to be in the range mentioned above, and for a typical value of η equal to 0,65, the data specifying the power plant are:

Δp= 785 Pa, Q= 80 m^3/s, P= 100 kW.

In this case we could use a 2,0m diameter fan specially designed for this purpose or 4 commercial fans of 1,0 m diameter, producing the same pressure increment, but with a volumetric flow of 20 m^3/s each. The latter option would reduce the total cost because the fans are a standard product.

4. GENERAL DESIGN PROCEDURE

The parameters that need to be defined in order to start the overall design are:

- Test chamber dimensions: width, W_{TC}, height, H_{TC}, and length, L_{TC}. These parameters allow to compute the cross-sectional area, $S_{TC}= W_{TC} H_{TC}$, and the hydraulic diameter, $D_{TC}=2 W_{TC}H_{TC}/(W_{TC}+ H_{TC})$.
- Contraction ratio, $N\approx5$ for low quality flow, and $N\approx9$ for high quality flow (considering the drawbacks of choosing a higher contraction ratio, explained before).
- Maximum operating speed, V_{TC}.

According to the impact on the wind tunnel dimensions and flow quality, Table 1 shows a classification of the design variables divided into two categories: main and secondary design parameters.

TABLE 1. Main and secondary wind tunnel design parameters

Main design parameters	Secondary design parameters
Maximum operating speed, V_{TC}	Contraction semi angle, $\alpha_C/2$
Test chamber width, W_{TC}	Settling chamber non-dimensional length, l_{SC}
Test chamber height, H_{TC}	Diffuser semi angle, $\alpha_D/2$
Test chamber length, L_{TC}	Diffuser 1 non-dimensional length, l_{D1}
Contraction ratio, N	Corner 1 expansion ratio, e_{C1}
	Corner 1 non-dimensional radius, r_{C1}
	Corner 4 non-dimensional radius, r_{C4}
	Diffuser 5 non-dimensional length, l_{D5}
	Corner 3 non-dimensional radius, r_{C3}
	Dimension of the fan matrix, n_W, n_H
	Unitary fan diameter, D_F
	Power plant non dimensional length, l_{PP}
	Corner 2 expansion ratio, e_{C2}

Now, following the guidelines given above, such as the convergence angle and the contour line shape of the contraction zone, the test and contraction chamber can be fully defined. In the case when both opening angles, α and β, are the same, the contraction length, L_C, is given by the expression:

$$L_C = \frac{\left(\sqrt{N} - 1\right) \cdot W_{TC}}{2 \cdot \tan\left(\alpha_C / 2\right)}.$$

Continuing in the upstream direction, the next part to be designed is the settling chamber. The only variable to be fixed is the length, because the section is identical to the wide section of the contraction. In the case when high quality flow is required, the minimum recommended non-dimensional length based on the hydraulic diameter, l_{SC}, is 0,60. This results from the necessity to provide extra space for the honeycomb and screens. In all other cases, the non-dimensional length may be 0,50. Therefore, the length of the settling, L_{SC}, chamber is given by:

$$L_{SC} = \sqrt{N} \cdot W_{TC} \cdot l_{SC}.$$

To obtain all the data for the geometric definition of the corner 4 satisfying all the recommendations given above we only need to fix the non-dimensional radius, r_{C4}. Its length, which is the same as its width, is:

$$L_{C4} = W_{C4} = \sqrt{N} \cdot W_{TC} \cdot (1 + r_{C4}).$$

Going downstream of the test chamber, we arrive at the diffuser 1. Assuming that both semi-opening angles are 3,5°, its non-dimensional length, l_{D1}, is the only design parameter. Although it has a direct effect on the wind tunnel overall length, we must be aware that this diffuser together with corner 1 are responsible for more than 50% of the total pressure losses. According to the experience, $l_{D1}>3$ and $l_{D1}>4$ is recommended for low and high contraction ratio wind tunnels respectively. The length of the diffuser 1, L_{D1}, and the width in the wide end, W_{WD1}, is defined by:

$$L_{D1} = W_{TC} \cdot l_{D1}$$

$$W_{WD1} = [1 + 2 \cdot l_{D1} \cdot \tan(\alpha_{D1}/2)] \cdot W_{TC}.$$

With regard to the corner 1, once its section at the entrance is fixed (it is constrained by the exit of diffuser 1), we must define the non-dimensional radius, r_{C1}, and the expansion ratio, e_{C1}. As a result, the width at the exit, W_{EC1}, the overall length, L_{C1}, and width, W_{C1}, can be calculated using:

$$W_{EC1} = W_{WD1} \cdot e_{C1}$$

$$L_{C1} = W_{WD1} \cdot (e_{C1} + r_{C1})$$

$$W_{C1} = W_{WD1} \cdot (1 + r_{C1}).$$

Therefore, we can already formulate the overall wind tunnel length, L_{WT}, as a function of the test chamber dimensions, the contraction ratio, and other secondary design parameters:

$$L_{WT} = L_{TC} + W_{TC} \cdot \left[\frac{(\sqrt{N} - 1)}{2 \cdot \tan(a_C/2)} + \sqrt{N} \cdot l_{SC} + \sqrt{N} \cdot (1 + r_{C4}) + l_{D1} + [1 + 2 \cdot l_{D1} \cdot \tan(a_{D1}/2)] \cdot (e_{C1} + r_{C1}) \right].$$

This quick calculation allows the designer to check whether the available length is sufficient to fit the wind tunnel.

Taking into account all the recommended values for the secondary design parameters, a guess value for the wind tunnel overall length, with a contraction ratio N=9 (high quality flow), is given by the formula:

$$L_{WT} = L_{TC} + 16 \cdot W_{TC}.$$

In the case when N=5 (low quality flow), the formula becomes:

$$L_{WT} = L_{TC} + 11{,}5 \cdot W_{TC}.$$

The designer must be aware that any modification introduced to the secondary design parameters modifies only slightly the factor that multiplies W_{TC} in the formulas above. Consequently, if the available space is insufficient, the only solution would be to modify the test chamber dimensions and/or the contraction ratio.

As we have already defined the wind tunnel length using the criterion of adequate flow quality, we can now devote our attention to designing the rest of the circuit, the so-called return circuit. The goal is not to increase its length, intending also to minimize the overall width and keeping the pressure loss as low as possible.

Keeping this in mind, the next step in the design is to make a first guess about the power plant dimensions. Following our design recommendations, a typical value for the total pressure loss coefficient of a low contraction ratio wind tunnel, excluding screens and honeycombs in the settling chamber, is 0,20, with respect to the dynamic pressure in the test chamber. This value is approximately 0,16 for a large contraction ratio wind tunnels. If screens and honeycombs were necessary, those figures could increase by about 20%.

As the power plant is placed more or less in the middle of the return duct, the area of the section will be similar to the mid-section of the contraction. Therefore, taking into account the volumetric flow, the total pressure loss, and the available fans, the decision about the type of fan and the number of them can be taken. Using this approach, the power plant would be defined, at least in the preliminary stage.

We will return now to the example we started before for the power plant section. To improve the understanding of the subject, we are going to present a case study. If the test chamber section was square and N=5, the mid-section of the contraction would be 1,67 x 1,67 m^2. This would allow us to place 4 standard fans of 0,800 m diameter each. The maximum reduction in the width size would be obtained by suppressing the diffuser 5, obtaining the wind tunnel platform shown in Figure 9. We have not defined the diffusion semi-angle in diffuser 3, but we checked afterwards that it was smaller than 3,5°. Figure 9 is just a wire scheme of the wind tunnel, made with an Excel spreadsheet, and for this reason the corners have not been rounded and are represented just as boxes.

In the case of a 4:3 ratio rectangular test chamber cross-section, the mid-section of the contraction would be 1,869x1,401 m^2 and for this reason we could suggest the use of 6 standard fans of 0,630 m diameter, organized in a 3x2 matrix, occupying a section of 1,890x1,260 m^2. Figure 10 shows the wire scheme of this new design. We can check that the diffuser 3 semi-angle is below 3,5° as well.

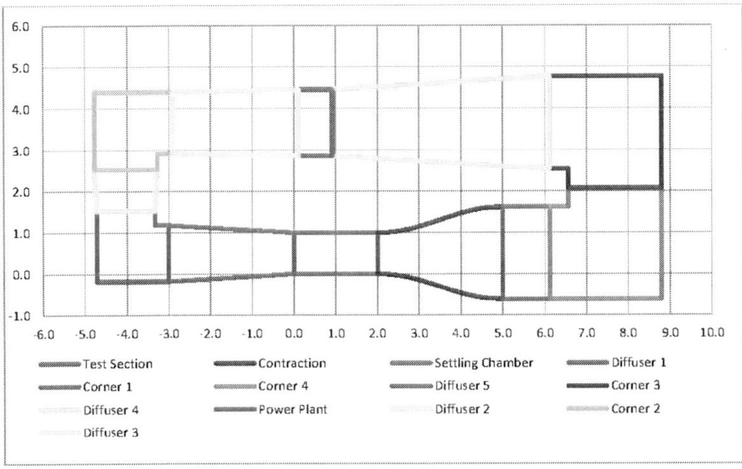

FIGURE 9. Non-dimensional scheme of a wind tunnel with square section test chamber and low contraction ratio, N≈5.

It is clear that the new design is slightly longer and wider, but it is because of the influence of the test chamber´s width, as shown above.

Notice that in both cases corner 3 has the same shape as corner 4. Similarly, the entrance section of diffuser 4 is the same as of the power plant section, and using a diffuser semi-angle of 3,5°, this item is also well defined.

At this stage we have completely defined the wind tunnel centre line, so that we can calculate the length, L_{CL}, and width, W_{CL}, using:

$$L_{CL} = (L_{C1} - W_{EC1}/2) + L_{D1} + L_{TC} + L_{C} + L_{ST} + (L_{C4} - W_{ED5}/2)$$

$$W_{CL} = (W_{C4} - W_{EC4}/2) + L_{D5} + (W_{C3} - W_{ED4}/2).$$

The distance between the exit of the corner 1 and the centre of the corner 2, DC1_CC2, can be calculated through the expression (see Figure 11):

$$D_{C1\_CC2} = W_{CL} - W_{ED1}\left(r_{C1} + \frac{1}{2}\right).$$

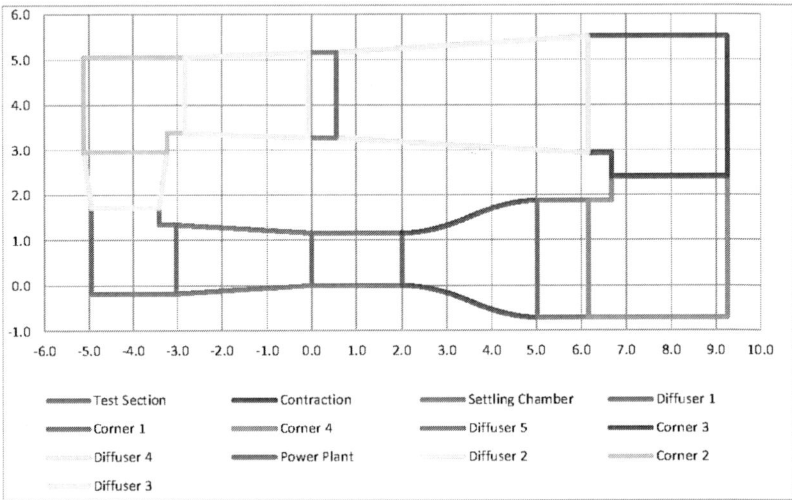

FIGURE 10. Non-dimensional scheme of a wind tunnel with rectangular section test chamber and low contraction ratio, N≈5.

On the other hand:

$$D_{C1\_CC2} = L_{D2} + [(W_{C2} - W_{EC2}/2)]$$

$$W_{EC2} = W_{ED2} \cdot e_{C2}$$

$$W_{C2} = W_{ED2} \cdot (r_{C2} + e_{C2})$$

$$W_{ED2} = W_{EC1} + 2 \cdot L_{D2} \cdot \tan(\alpha_{D2}/2).$$

Manipulating and combining those equations, we obtain:

$$L_{D2} = \frac{D_{C1\_CC2} - W_{EC1} \cdot (r_{C2} + e_{C2}/2)}{1 + 2 \cdot (r_{C2} + e_{C2}/2) \cdot \tan(\alpha_{D2}/2)}.$$

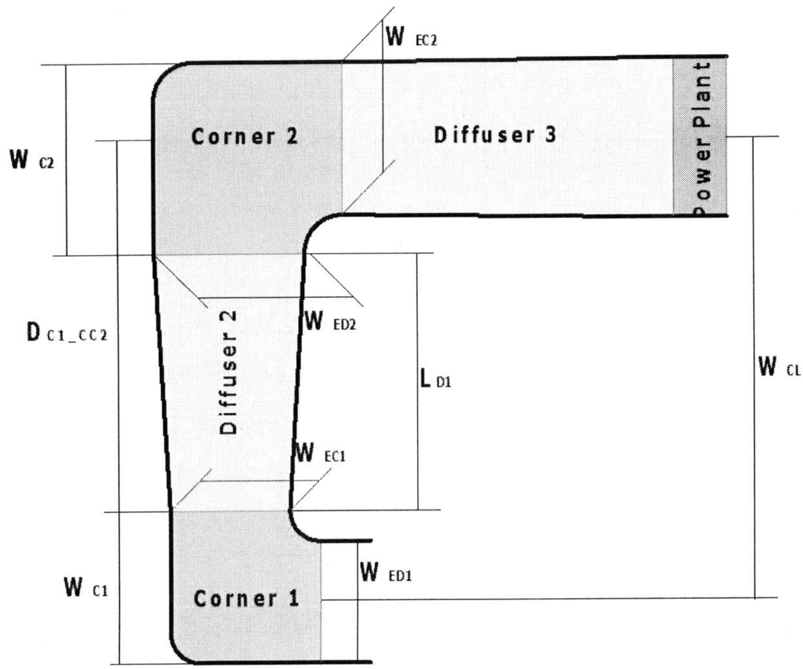

FIGURE 11. Scheme with the definition of the variable involving the design of diffuser 2 and 3, and corner 2.

With this value, by substituting it into the previous expressions, we have all the parameters to design diffusers 2 and 3, and corner 2. Finally, it is necessary to check that the opening angles of diffuser 3 are below the limit. In case when the vertical opening angle, α, exceeds the limit, the best option is to increase the diffuser 1 length, if this is possible, because it improves flow quality and reduces pressure loss. If the wind tunnel length is in the limit, another option is to add the diffuser 5 to the original scheme. However, it will increase the overall width. When the limit of the horizontal opening angle, β, is exceeded, then the best option is to adjust the values of the expansion ratio in corners 1 and 2, because it will not change the overall dimensions.

The following case study is a wind tunnel with high contraction ratio, $N \approx 9$, and square section test chamber. In this case, the approximate area of the power plant section will be 2,000 x 2,000 m^2. In this case we have two compatible options to select the power plant. We can just select a matrix of 4 fans, 1,000 m diameter each. However, if the operating speed is rather high, in order to be able achieve the required pressure increment and the mass flow, we may need to use 1,250 m diameter fans. Figure 12shows both options. Note that the overall planform is only slightly modified and the only difference is the position where the power plant is placed.

The design of the diffusers 2 and 3, and the corner 2 will be done following the same method as for the previous cases.

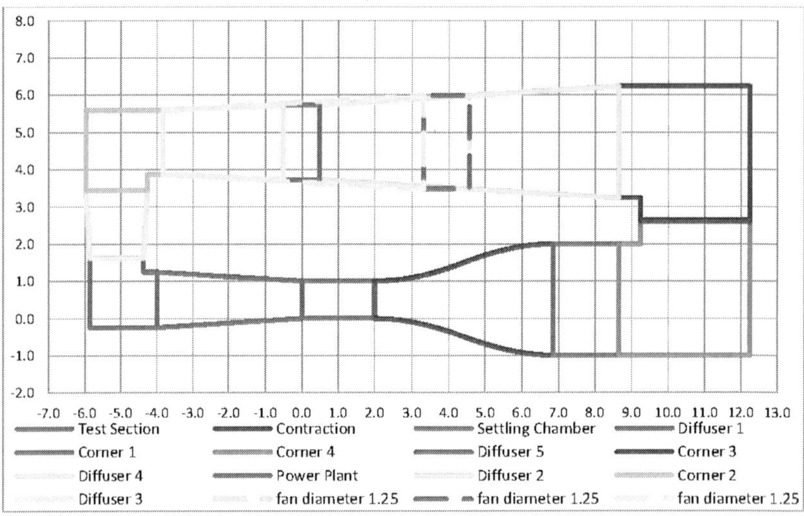

FIGURE 12. Non-dimensional scheme of a wind tunnel with square section test chamber and high contraction ratio, N≈9. Two different standard power plant options are presented.

5. WIND TUNNEL CONSTRUCTION

One of the most important points mentioned in this chapter refers to the wind tunnel cost, intending to offer low cost design solutions. Up to now we have mentioned such modifications to the power plant, proposing a multi-fan solution instead of the traditional special purpose single fan.

The second and most important point is the wind tunnel's construction. The most common wind tunnels, including those with square or rectangular test sections, have rounded return circuits, like in the case of the NLR-LSWT. However, the return circuit of DNW wind tunnel is constructed with octagonal sections. Although the second solution is cheaper, in both cases different parts of the circuit needed to be built in factories far away from the wind tunnel location, resulting in very complicated transportation operation.

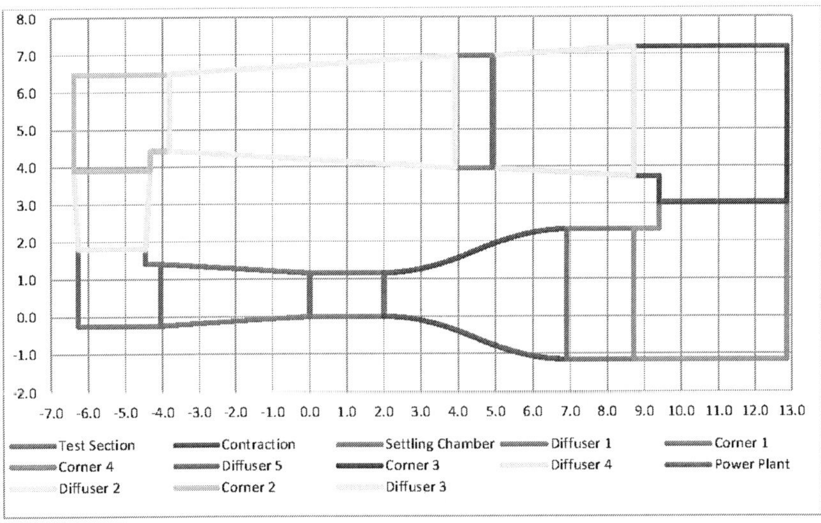

FIGURE 13. Non-dimensional scheme of a wind tunnel with rectangular section test chamber and large contraction ratio, N≈9.

To reduce the costs, all the walls can be constructed with flat panels, which can be made on site from wood, metal or even concrete, like in the case of ITER's wind tunnel. Figure 14 shows two wind tunnels built with wood panels and aluminium standard profile structure.

Both wind tunnels shown in Figure 14 are open circuit. The one on the left is located in the Technological Centre of the UPM in Getafe (Madrid) and its test chamber section is 1,20 x 1,00 m^2. Its main application is mainly research. The right one is located in the Airplane Laboratory of the Aeronautic School of the UPM. Its test chamber section is 0,80 x 1,20 m^2, and it is normally used for teaching purposes, although some research projects and students competitions were done there as well. Despite the fact that these tunnels are open circuit, the construction solutions can be also applied to closed circuit ones.

FIGURE 14. Research and education purpose wind tunnels built with wood panel and standard metallic profiles, with multi-fan power plant.

According to our experience, the manpower cost to build a wind tunnel like those defined in figures 9 to13 could be 3 man-months for the design and 16 man-months for the construction. With these data, the cost of the complete circuit, excluding power plant, would be about 70.000,00 €. In our opinion, the cost figure is very good, considering the fact that the complete building time possibly may not exceed even 9 months.

We have more reliable data with regard to the ITER wind tunnel, built in 2000-01. The whole cost of the wind tunnel, including power plant, work shop and control room, was 150.000,00 €.

This wind tunnel was almost completely built with concrete. Figure 15 shows different stages of the construction, starting from laying the foundations to the almost final view. The small photos show the contraction, with the template used for wall finishing, and the power plant.

6. CONCLUSIONS

A method for quick design of low speed and low cost wind tunnels, either for aeronautical and/or civil applications, has been presented.

The possibility of deciding between both applications means that the method allows achieving flow quality level as good as desired.

The method also allows to the designer to get a quick and rough estimation of the overall wind tunnel size, once the main design parameters are given.

FIGURE 15. Photographic sequence of the construction of the ITER Low Speed Wind Tunnel. The top left picture shows the foundations, the top right the contraction, the bottom left the power plant and the bottom right a view from the outside almost at the end of the construction.

The guidelines to choose the secondary design parameters are given as well.

To address the low cost of design and construction, the use of a multi fan power plant and rectangular duct sections is proposed as well.

NOMENCLATURE

a_i, b_i, c_i, d_i	Family of polynomial coefficients of the contraction contour shape	
CFD	Computational Fluid Dynamics	
D_{C1_CC2}	Distance between the exit of the corner 1 and the centre of the corner 2	m
D_F	Unitary fan diameter	m
D_H	Studied duct section hydraulic diameter	m
e_i	Corner 'i' expansion ratio	
F_0	Area of the diffuser's narrow section	m^2
F_1	Area of the diffuser's wide section	m^2
H_{ent}	Section height of the duct´s entrance	m
H_{exit}	Section height of the duct´s exit	m

L	Studied duct length	m
L_i	Duct 'i' length	m
l_i	Duct 'i' non-dimensional length	
LSWT	Low Speed Wind Tunnel	
L_{WT}	Overall wind tunnel length	m
N	Contraction ratio	
n	Average number of corner vanes	
n_W, n_H	Dimensions of the fan matrix	
P	Power of the power plant	W
Q	Volumetric flow	m^3/s
r	Corner radius	m
Re	Reynolds number based on the hydraulic diameter	
r_i	Corner 'i' non-dimensional radius	

S	Diagonal dimension of the corner	m
t_1	Chord of the corner vane	m
V	Operating speed at the test chamber	m /s
V_{TC}	Maximum operating speed at the test chamber	m /s
W_{CL}, L_{CL}	Wind tunnel centre line width and length	m
W_{ent}	Section width of the duct´s entrance	m
W_{exit}	Section width of the duct´s exit	m
W_{ij}, H_{ij}	Duct j width and height of the i section (wide-end, $_W$; narrow-end, $_N$; constant,)	m
WTT	Wind Tunnel Tests	
(x_N, y_N)	Narrow-end coordinates of the contraction contour shape	
(x_W, y_W)	Wide-end coordinates of the contraction contour shape	
$\alpha_i/2$	Vertical contraction/opening semi-angle of	de

	the duct 'i'	g
$\beta_i/2$	Horizontal contraction/opening semi-angle of the duct 'i'	deg
Δp	Pressure increment at the power plant section	Pa
$\zeta\zeta$	Total pressure loss coefficient	
$\zeta f\zeta f$	Friction pressure loss coefficient	
$\zeta M\zeta M$	Singular pressure loss coefficient of a corner	
$\eta\eta$	Fan efficiency	
$\lambda\lambda$	Friction coefficient per non-dimensional length of the studied duct	
$\rho\rho$	Operating air density	kg/m^3

REFERENCES

1. J. B Barlow, W. H Rae, A Pope, Low-speed wind tunnel testing, John Wiley & Sons New York, 1999rd ed.

2. G. G Borger, The optimization of wind tunnel contractions for the subsonic range, NASA Technical Translation / F-16899, NASA Washington, 1976

3. W. T Eckert, K. W Mort, J Jope, Aerodynamic design guidelines and computer program for estimation of subsonic wind tunnel performance, NASA technical note / D-8243, NASA Washington, 1976

4. S. M Gorlin, I. I Slezinger, Wind tunnels and their instrumentation, Israel Program for Scientific Translations Jerusalem, 1966

5. Idel′Cik. I.E., Memento des pertes de charge: Coefficients de pertes de charge singulières et de pertes de charge par frottement, Eyrolles Editeur, Paris 1969

6. E. C Maskell, A theory of the blockage effects on bluff bodies and stalled wings in a closed wind tunnel, R. & M. 3400, November, 1963

7. R. D Mehta, and P Bradshaw, Design Rules for Small Low-Speed Wind Tunnels, Aero. Journal (Royal Aeronautical Society), 734431979

8. J Scheiman, Considerations for the installation of honeycomb and screens to reduce wind-tunnel turbulence, NASA Technical Memorandum / 81868, NASA Washington, 1981

9. The Royal Aeronautical Society. Wind tunnels and wind tunnel test techniques, Royal Aeronautical Society London, 1992

4

CHAPTER

Critical embedment length and bond strength of fully encapsulated rebar rockbolts

Charlie C. Li, Gunnar Kristjansson, Are Håvard Høien

Department of Geology and Mineral Resources Engineering, The Norwegian University of Science and Technology (NTNU), Norway

ABSTRACT

A series of rock bolt pull tests were carried out in the laboratory to determine the critical embedment length of a specific type of fully cement-grouted rebar bolt. The rebar bolt is 20 mm in diameter, and it is widely used in underground excavations in Norway. Three water-cement (w/c) ratios were used in the tests. It was discovered that the critical embedment length of the rock bolts was approximately 25 cm for the water-cement ratio 0.40 (the corresponding uniaxial compressive strength (UCS) of the grout is 37 MPa), 32 cm for the ratio 0.46 (UCS 32 MPa), and 36 cm for the ratio 0.50 (UCS 28 MPa), for the specific type of cement, Rescon zinc rock bolt cement. It was found that the bond strength of the rock bolt is not a constant but is related to the embedment length. The bond strength was linearly proportional to the UCS of the grout.

Keywords: Rockbolt; Embedment length; Pullout test; Bond strength

1. INTRODUCTION

The rebar rock bolt that is fully encapsulated in a borehole with cement mortar or resin grout is the type of rock bolt widely used in civil and mining engineering. The wide use of this type of rock bolt is predominately owed to its high load-bearing capacity. The grouting quality and performance of the rock bolt are usually examined by pull test. In such a test, a rock bolt grouted in a borehole is either pulled out to measure the bond strength in the case of a short embedment length (Hyett et al., 1995, Benmokrane et al., 1992, Benmokrane et al., 1996 and Zhao and Yang, 2011) or pulled until failure of the rock bolt shank (Stillborg, 1994, Stjern, 1995 and Chen and Li, 2015). Pull tests may be conducted in the field (Franklin and Woodfield, 1971, Franklin et al., 1974, Bjurholt, 2007 and Soni, 2000) or in laboratories (Ito et al., 2001 and Li et al., 2014).

The performance of a fully grouted rebar rock bolt is very much affected by the bond at the rock bolt-grout interface. The bond strength directly determines the critical embedment length of the rock bolt, which refers to the longest encapsulated length at which the rock bolt is pulled out of the borehole. The rock bolt will fail in the shank if the grouted section is longer than the critical embedment length. For instance, in order to stabilise a loosened rock block with fully grouted rock bolts, the length of the rock bolt portion that is installed in the stable stratum must be longer than the critical embedment length. The quality of grouting and the load-bearing capacity of grouted rock bolts are often examined through pull tests in the field. One of the concerns with such a pull test is whether the performance of the rock bolt is negatively affected by the pull test. The authors acknowledge that the issue of the critical embedment length was studied by others in conjunction with individual engineering projects, but little information about it can be found in the literature. Therefore, a series of tests were carried out at the Laboratory of Rock Mechanics at the Norwegian University of Science and Technology to determine the critical embedment length of a specific type of 20 mm rebar rock bolt using the cement mortar Rescon zinc rock bolt cement mix in three different water-cement ratios. Both the rock bolt and the cement mix are widely used for rock support in Norwegian tunnels and underground caverns. The test results are reported in this article. In addition to the determination of the critical embedment length, the bond strength of the fully grouted rock bolts was also studied. It was found that the bond strength is not a constant but is related to the embedment length.

2. SPECIMENS

2.1. Rock bolts

The type of rebar rock bolt tested is 20 mm in diameter, which is widely used in road tunnels in Norway (Fig. 1). It is made of steel B500NC according to the Norwegian standard NS 3576-3, which specifies the mechanical properties of the steel, the formation of the rib pattern, and the production control. For the sake of corrosion protection, the rock bolt surface is treated by hot zinc-galvanisation with a minimum thickness of 65 μm and a coating of epoxy powder with a minimum thickness of 60 μm. The characteristic yield and ultimate tensile loads of the bolt are 157 kN and 188 kN, respectively.

2.2. Cement grout

The grouting agent used in the tests is Rescon zinc rock bolt cement mix, which is used for grouting rock bolts in road tunnels in Norway. The cement mix is made of cement (c) and silica (s). The weight ratio of the mix (cement and silica) to the cement is $(c + s)/c = 1.7$. The ratio of the mix $(c + s)$ to the water (w) has the following relationship with the conventional water-cement ratio (w/c):

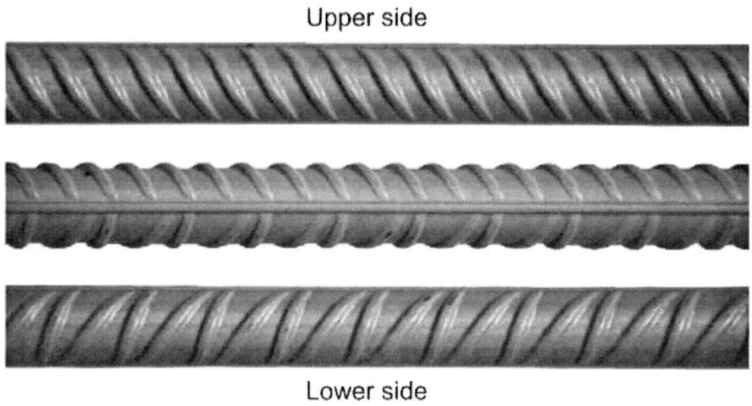

Fig. 1. The rib pattern of the rebar bolt, viewed in three orthogonal directions. Bolt diameter 20 mm, rib spacing approximately 12.5 mm.

$$\frac{w}{c+s}=\frac{c}{c+s}\frac{w}{c}=\frac{1}{1.7}\frac{w}{c} \tag{1}$$

For instance, the ratio of w/(c + s) should be 0.235 if the water-cement ratio of the grout is intended to be 0.40. Three water-cement ratios, 0.40, 0.46, and 0.50, were used in the tests and the corresponding water-mix ratios (i.e., w/(c + s)) were 0.235, 0.270, and 0.294, respectively.

3. TESTING

3.1. Test plan

The embedment length of the rock bolt was varied for every water-cement ratio in order to find the critical embedment length. In general, three rock bolt specimens were pulled for every embedment length. The number of rock bolt specimens for every embedment length and every water-cement ratio is given in Table 1.

Table 1. Number of rock bolt specimens for every water-cement ratio and for every embedment length.

Embedment length (cm)	Water-cement ratio		
	0.40	0.46	0.50
10	3	3	3
15	3	–	–
20	3	3	3
25	–	3	–
30	3	3	3
40	1	–	2
Total	13	12	11

The uniaxial compressive strength (UCS) of the grout was measured after the same curing time as the tested rock bolts. Three cubic grout specimens, 100 × 100 × 100 mm in size, were prepared when the rock bolt was grouted and then tested on a servo-controlled test machine, GCTS RTR-4000, on the same day the rock bolt was tested.

3.2. Test arrangement

Boreholes were percussively drilled with a 48 mm drill bit in a cubic concrete block with a dimension of 950 × 950 × 950 mm. The UCS of the concrete is approximately 110 MPa. The grout was mixed with one of the water-cement ratios given in Table 1. The ready-mixed grout was pumped into the hole and the rock bolt was inserted to a desired depth. The curing time was scheduled to be seven days, but some of the rock bolts were pulled out after eight days. When a rock bolt was tested, a cylindrical spacer, collaring the rock bolt, was placed on the top of the concrete block, a hydraulic cylinder (jack) was placed on the top of the spacer, and finally a rock bolt plate and a barrel-and-wedge unit were placed on the top of the setup string to fasten the rock bolt (Fig. 2). The purpose of using the barrel-and-wedge to replace the nut and thread is to avoid premature failure of the rock bolt in the thread. With such an arrangement, the rock bolt does not need to be threaded so that the rock bolt head is not weakened due to threading. Thus, the failure occurs either in the grout along the rock bolt-grout interface or in the bolt steel under the pull load. The distance from the surface of the concrete block to the barrel-and-wedge, that is, the freely stretched length of the rock bolt is 60 cm. The movement of the bolt head was measured with respect to the base of the hydraulic cylinder with an extensometer. Rock bolts with long enough embedment lengths will fail in the shank. To avoid damage to the extensometer after the violent failure of the bolt shank, when the rock bolt risks failure in the steel, the test was terminated when the load reached the tensile strength of the rock bolt but "necking" had not started.

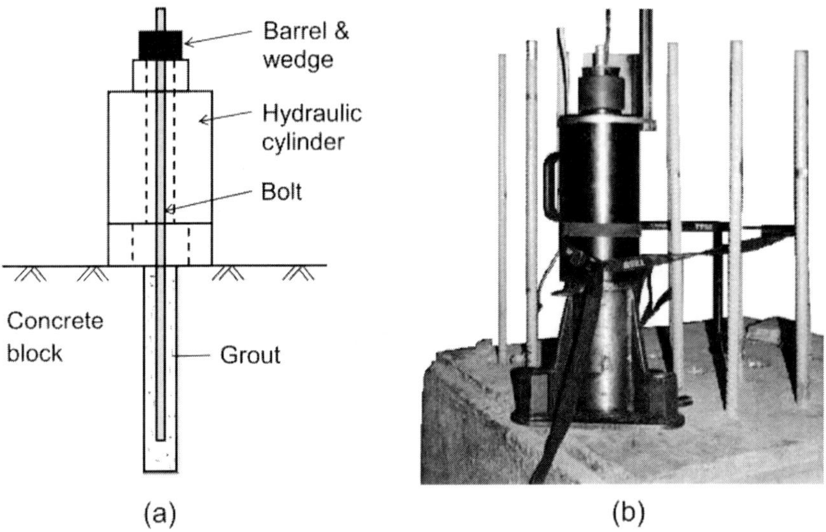

Fig. 2. Pull test arrangement. (a) A sketch and (b) the test arrangement.

4. RESULTS

4.1. Pull test results

Every rock bolt specimen is labelled with three digits after the letter B. The first digit represents the parameter of water-cement ratio, the second represents the group of the embedment length, and the third represents the sequence number of the specimen in the group. The rock bolts were tested after curing for seven to nine days.

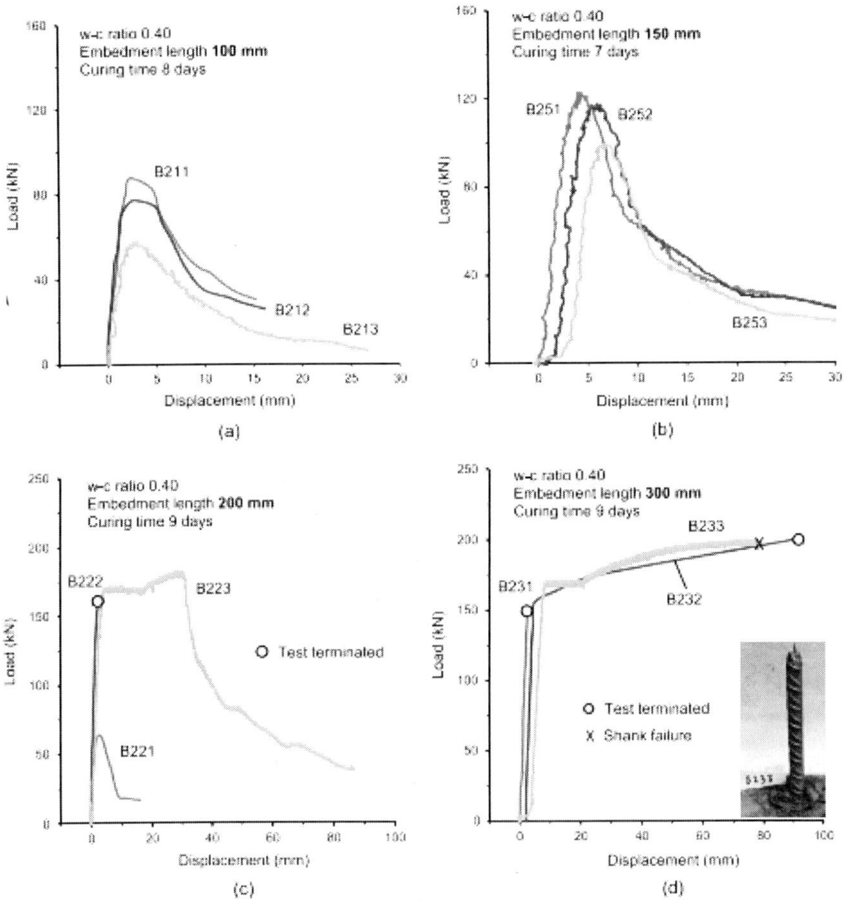

Fig. 3. Pull load–displacement curves of the bolts grouted with water-cement ratio 0.40 for embedment lengths (a) 100 mm, (b) 150 mm, (c) 200 mm, and (d) 300 mm.

4.1.1. Water-cement ratio 0.40

The load-displacement curves of the rock bolts grouted with a water-cement ratio of 0.40 are presented in Fig. 3. The test results of rock bolts with an embedment length of 100 mm are shown in Fig. 3a. The displacement of rock bolt B213 was measured using an extensometer, but the other two rock bolts, B211 and B212, were measured manually using a calliper. The ultimate load was 57 kN for B213 but 77 and 87 kN, respectively, for B211 and B212. All three rock bolts slipped in the grout to the end.

The test results of rock bolts with an embedment length of 150 mm are shown in Fig. 3b. The ultimate load varied from 97 to 118 kN, and all three rock bolts in the group slipped along the bolt-grout interface. Rock bolt B253 had the lowest load at 97 kN. It was observed that the borehole was not fully cement filled to the borehole collar so that its actual embedment length was a little shorter than 15 cm. Thus, it was excluded in the calculation of the average bond strength of the rock bolts.

The test results of rock bolts with an embedment length of 200 mm are shown in Fig. 3c. The ultimate load was only 64 kN for rock bolt B221. The low ultimate load might be owing to poor grouting. The test of B222 was wrongly interrupted when the load was at a level close to the yield load. Therefore, only the test result of bolt B223 is representative in this group. Bolt B223 first yielded and elongated up to 30 mm, and then the embedded section started to slip at a load of 180 kN.

The test results of rock bolts with an embedment length of 300 mm are shown in Fig. 3d. The tests of both B231 and B232 were terminated before the rupture of the bolt shank. The test of bolt B231 was terminated just prior to yielding. Bolt B232 yielded and displaced 90 mm when the test was terminated. It was expected that it would fail in the shank if the loading continued. Bolt B233 failed in the shank after a displacement of 78 mm. The insert picture in the figure shows the necking failure of the bolt. No slippage occurred for all three rock bolts.

4.1.2. Water-cement ratio 0.46

The load-displacement curves of the rock bolts with a water-cement ratio of 0.46 are presented in Fig. 4. The test results of rock bolts with an embedment length of 100 mm are shown in Fig. 4a. All three rock bolts in the group slipped in the grout, and their ultimate loads varied from 39 to 48 kN.

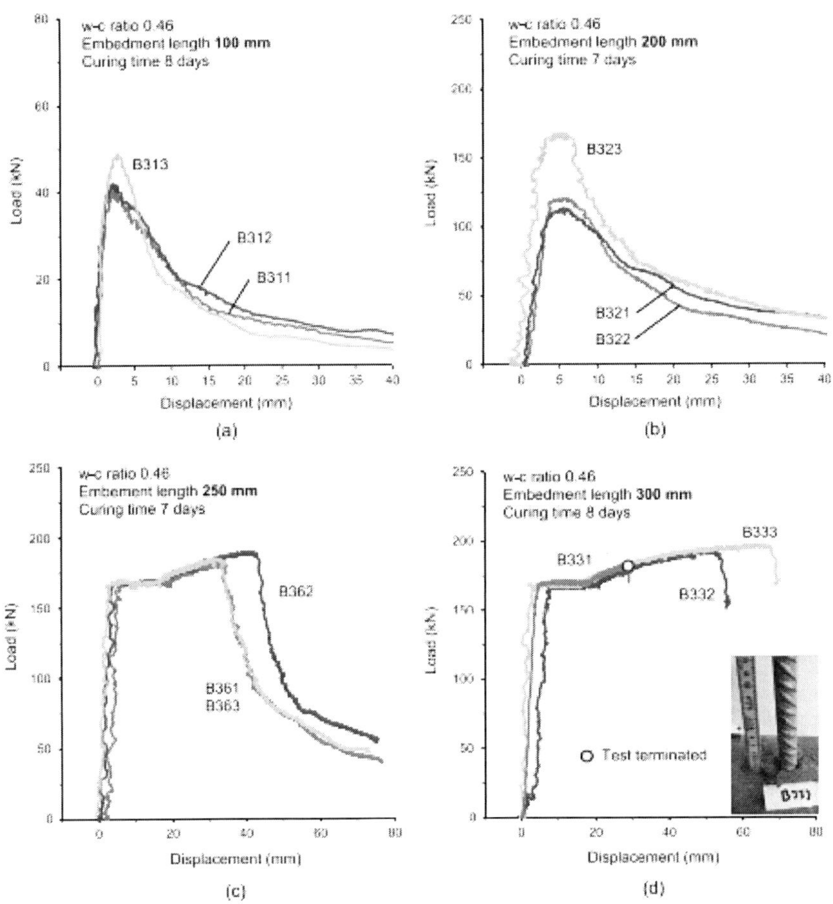

Fig. 4. Pull load–displacement curves of the bolts grouted with water-cement ratio 0.46 for embedment lengths (a) 100 mm, (b) 200 mm, (c) 250 mm, and (d) 300 mm.

The test results of rock bolts with an embedment length of 200 mm are shown in Fig. 4b. All the three rock bolts in the group slipped after the peak loads. The ultimate loads of B321 and B322 were similar, 113 and 119 kN, respectively, but the ultimate load of B323 was significantly higher at 167 kN. It was found after the test that the embedment length of B323 was approximately 230–240 mm, longer than the other two rock bolts. The result of B323 is thus excluded in the calculation of the average bond strength of the bolts in the group.

The test results of rock bolts with an embedment length 25 cm are shown in Fig. 4c. All three rock bolts in the group yielded and started to slip in the grout at 180 or 190 kN.

The test results of rock bolts with an embedment length of 300 mm are shown in Fig. 4d. All three rock bolts in the group yielded, and rock bolts B332 and B333 started to slip in the grout at approximately 200 kN. The test of B331 was terminated before slippage. The embedment length of 300 mm is very close to the critical length so that no embedment lengths longer than 300 mm were tested for the water-to-cement ratio of 0.46.

4.1.3. Water-cement ratio 0.50

The load-displacement curves of the rock bolts with a water-cement ratio of 0.50 are presented in Fig. 5. The test results of rock bolts are shown in Fig. 5a for those with an embedment length of 100 mm and in Fig. 5b for those with an embedment length of 200 mm. All rock bolts in these two groups slipped in the grout at a load below the yield load of the bolt shank.

The test results of rock bolts with an embedment length of 300 mm are shown in Fig. 5c. The shanks of all three rock bolts in the group yielded first and then slipped in the grout.

The test results of rock bolts with an embedment length of 400 mm are shown in Fig. 5d. The bolts did not show any sign of slippage in the grout so that the tests were terminated when the loads reached 200 kN. Clearly, the critical embedment length of the bolts is shorter than 400 mm for this water-cement ratio.

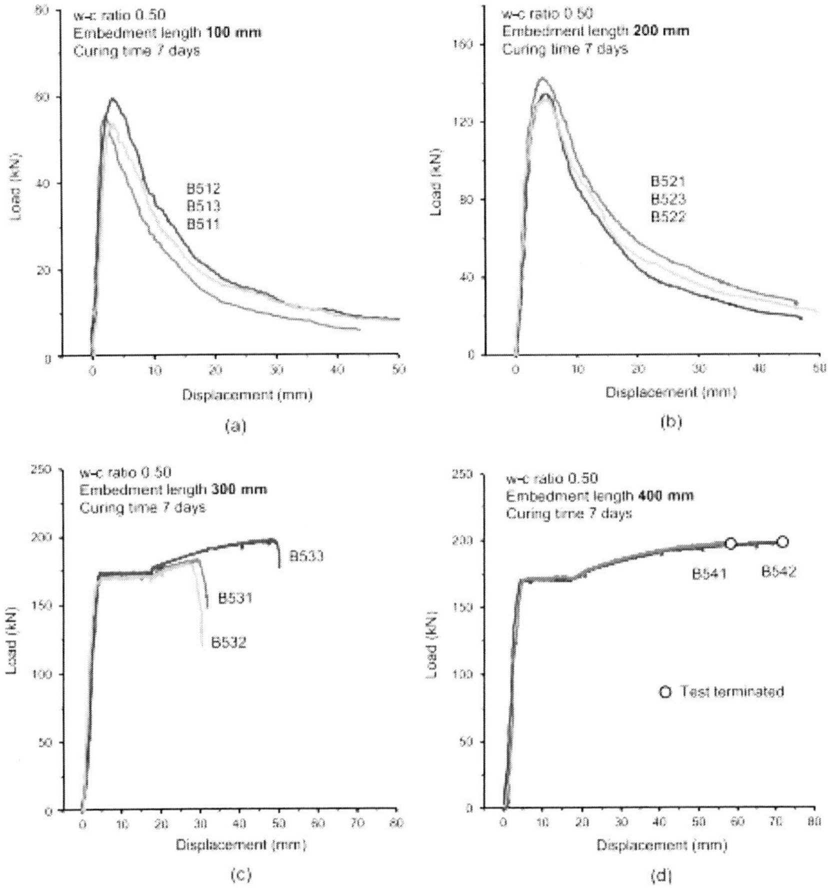

Fig. 5. Pull load–displacement curves of the bolts grouted with water-cement ratio 0.50 for embedment lengths (a) 100 mm, (b) 200 mm, (c) 300 mm, and (d) 400 mm.

4.2. UCS results of the grout cubes

The cubic specimens of the grout for every water-cement ratio were tested for UCS after seven-eight days of curing. Abnormal results were obtained for the water-cement ratio 0.50 in the first test round so that a second round of tests was run afterward. The UCS results of the grouts for different water-cement ratios are presented in Table

2 and in Fig. 6. Note that the abnormal UCS values for the water-cement ratio 0.50 in the first round were excluded in the figure.

Table 2. UCS values of the cubic specimens of the grouts after 7–8 days of curing (Specimen size: 100 × 100 × 100 mm).

w/c ratio	UCS (MPa)	
	Round 1	Round 2
0.40	33, 49, 43	38, 36, 37
0.46	29, 31, 29	33, 32, 32
0.50	45[a], 43[a], 43[a]	27, 28, 30

[a] Abnormal values.

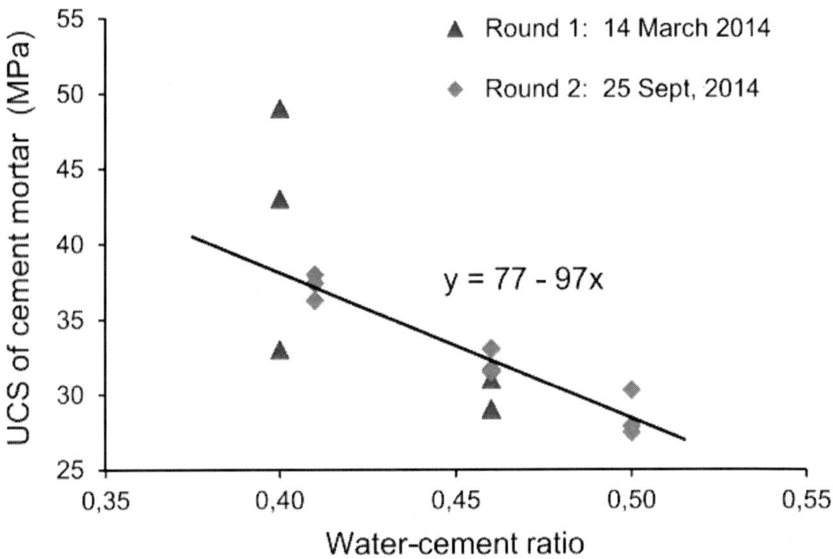

Fig. 6. UCS values of the grouts, measured with cubic specimens, for different water-cement ratios after 7-day curing.

5. ANALYSIS OF THE TEST RESULTS

5.1. The critical embedment length

The ultimate pull loads of the bolt specimens are plotted against the embedment lengths for the three different water-cement ratios in Fig. 7, Fig. 8 and Fig. 9. The critical embedment length of the rebar rock bolt for a given water-cement ratio is determined so that the points representing slippage fit a regression curve and the abscissa of the intersection of the regression curve with the strength line (at 200 kN) of the rock bolt shank is defined as the critical embedment length. Fig. 7 is the load-embedment length diagram of the rock bolts for the water-cement ratio 0.40. Notice that the results of rock bolts B253, B221, B222, and B231 were excluded in the plot for the reasons described in Section 4.1. The specimen with a 300 mm embedment length failed in the shank, while the others slipped under their ultimate loads, indicating bond failure at the rock bolt-grout interface. The critical embedment length of the rock bolt for the water-cement ratio 0.40 is determined to be approximately 250 mm.

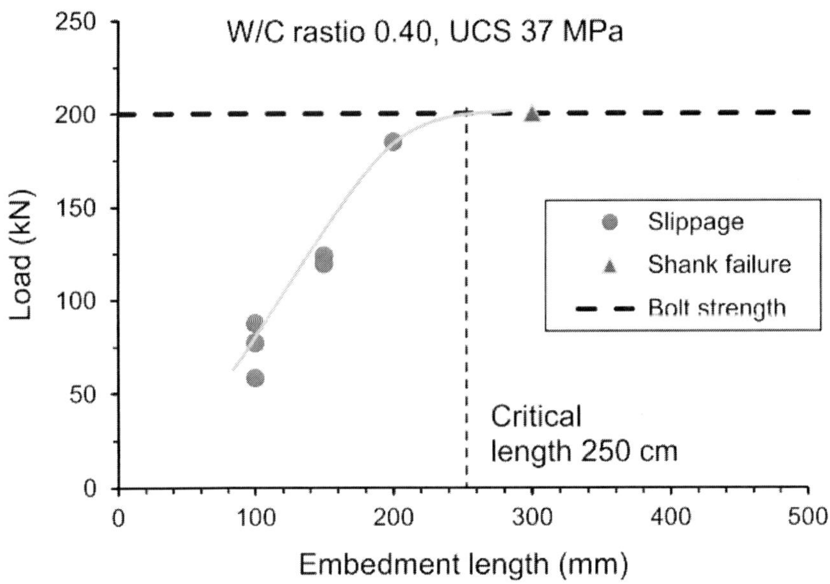

Fig. 7. Ultimate pull load versus the embedment length of the bolt for water-cement ratio 0.40.

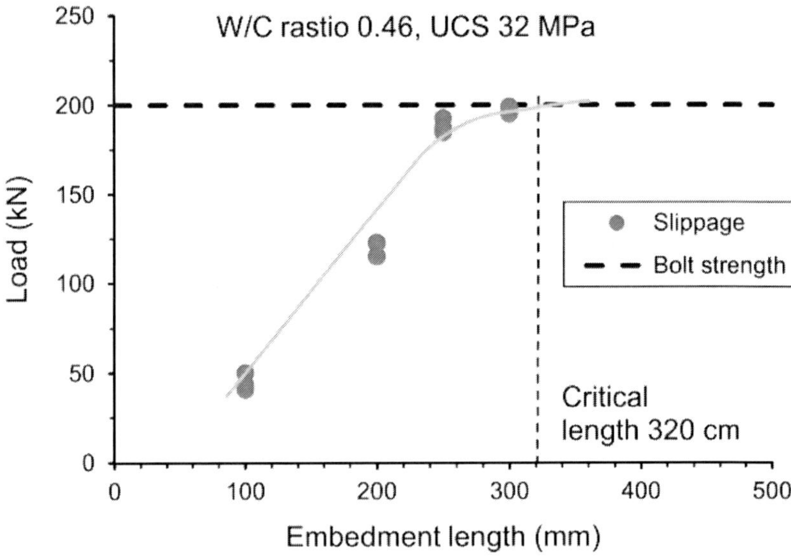

Fig. 8. Ultimate pull load versus the embedment length of the bolt for water-cement ratio 0.46.

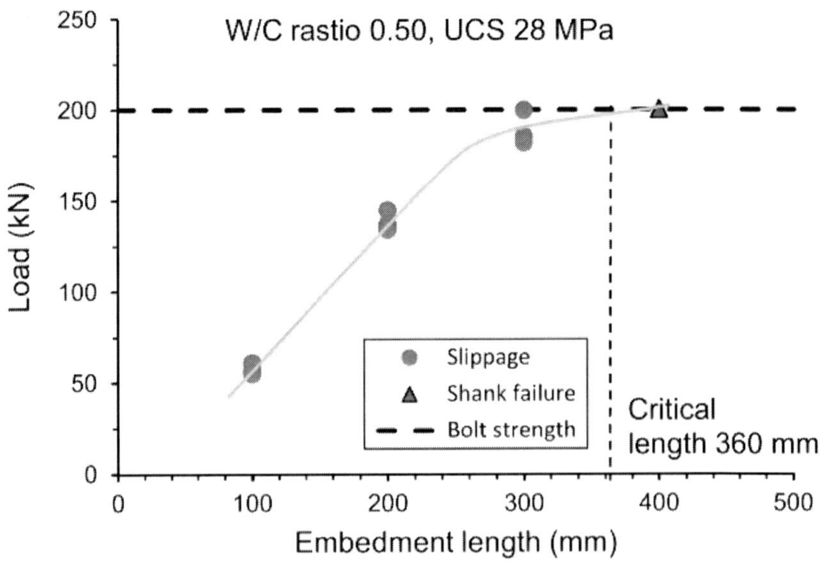

Fig. 9. Ultimate pull load versus the embedment length of the bolt for water-cement ratio 0.50.

The ultimate loads of the rock bolt specimens in the group with a water-cement ratio of 0.46 are plotted against the embedment length in Fig. 8. Notice that the result of rock bolt B323 was excluded in the plot because of its abnormality. All the rock bolts slipped in the grout, but the rock bolts with a 300 mm embedment length started to slip when the load was at a level close to the ultimate load of the bolt shank (see Fig. 4d), indicating that this length is close to the critical embedment length. The critical embedment length of the rock bolt for the water-cement ratio 0.46 is determined to be approximately 320 mm.

All the rock bolts with a water-cement ratio of 0.50, except those with an embedment length of 400 mm, slipped in the grout (Fig. 9). The two rock bolts with an embedment length of 400 mm did not slip when the tests were terminated at load levels very close to the tensile strength of the bolt to avoid damage to the measurement instrument (seeFig. 5d). Thus, these two rock bolts were treated as failures in the diagram. The critical embedment length of the rock bolt for the water-cement ratio 0.50 is determined to be approximately 360 mm.

The critical embedment length of the bolt, L_C, is linearly related to the water-cement ratio, as shown in Fig. 10, as follows:

$$L_C = 110 \times (w/c) - 19.1 \tag{2}$$

and to the UCS of the grout as follows:

$$L_C = 72.5 - 1.28 \times UCS. \tag{3}$$

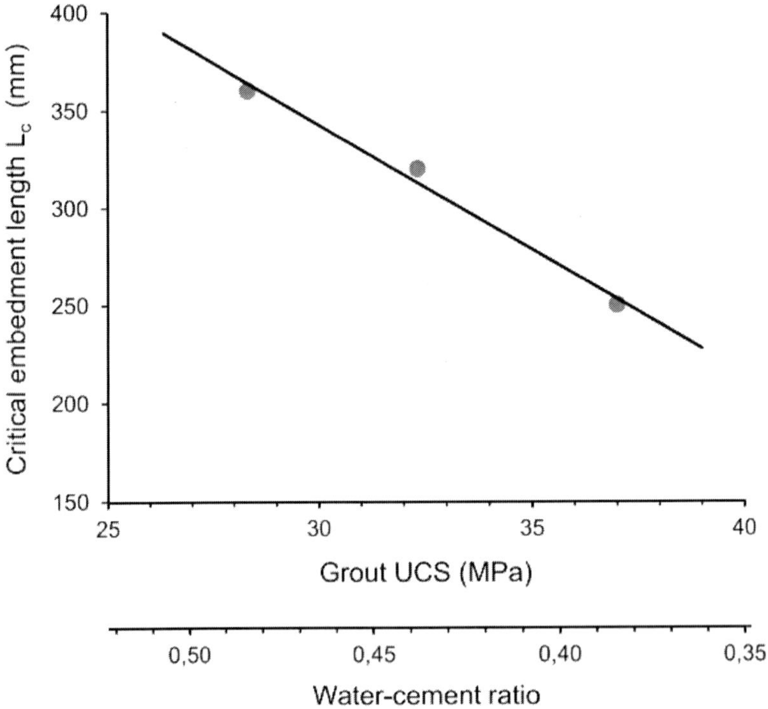

Fig. 10. Critical embedment length versus the UCS of grout and the water-cement ratio.

The critical embedment lengths derived in this section were based on the results of the pull tests carried out under strictly controlled laboratory conditions, which guaranteed satisfactory grouting qualities. The critical embedment length in the field could be longer than the laboratory-obtained value, taking into account variations in grouting quality as well as rock mass quality. Therefore, a safety factor between two and four should be used for the critical anchoring length of rock bolts in the support design (Littlejohn, 1992).

5.2. Bond strength

Bond strength, defined as the average shear strength, of the rock bolt-grout interface, is usually used to describe the binding capacity of an encapsulated rock bolt with the grout. It is calculated as:

$$s = P_{max}/\pi d_b l_o, \tag{4}$$

where P_{max} is the ultimate pull load in the case of bolt slippage in the grout, d_b is the rock bolt diameter, and l_o is the embedment length of the rock bolt. It makes more sense to plot the bond strength against the UCS of grout than the cement-water ratio because the strength of the grout is not only dependent on the cement-water ratio but also on the cement type. The average bond strengths of the rock bolts for three different grouts of UCS 32, 37, and 47 MPa are calculated according to Eq. (4) and are plotted in Fig. 11. The data for the grout of UCS 47 MPa are taken from another series of pull tests, which are plotted in the figure for the sake of comparison. The last point for UCS 32 MPa grout, marked by a circle in the figure, is the value for the rock bolts the shanks of which significantly yielded at a strain of approximately 10%. Except that point, the bond strength tends to increase with the embedment length of the rock bolt for all three types of grouts. Note that the grouting cement mortar failed in the collar of the borehole, forming a crater (see the photos in Figs. 3d and 4d). Thus, the average bond strength should be slightly higher than that presented in Fig. 11 if the effective embedment length is used for the calculation.

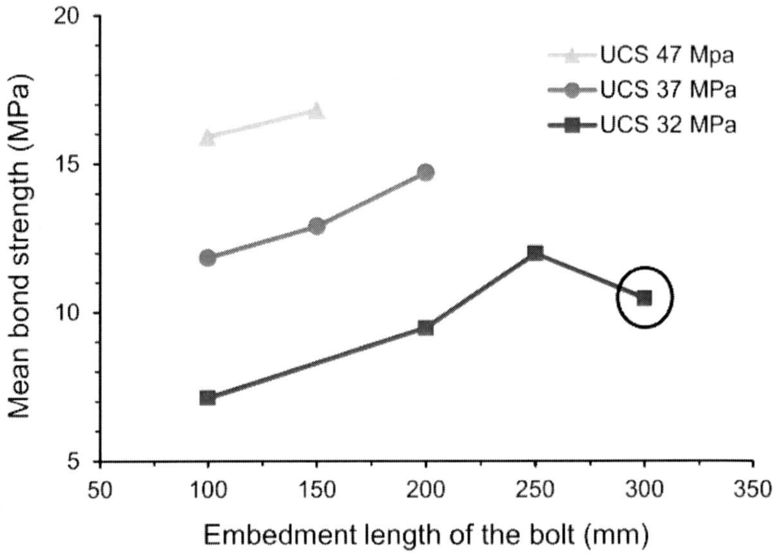

Fig. 11. The average bond strength versus the embedment length of the bolt. The circled point represents the specimen the shank of which was significantly yielded (approximately 10% strain).

It is noticed that Kilic et al. (2002) experimentally investigated the influence of the embedment length on bond strength. In their laboratory tests, the rebar bolts were 12 mm in diameter, installed in 22 mm holes with Portland cement mortar of UCS 35.5 MPa (with a water-cement ratio of 0.4). The embedment length varied from 15 to 32 cm. The bond strengths of the bolts varied in a very small interval from 7.58 to 7.98 MPa. In other words, their tests showed that the bond strength is independent of the embedment length. Further experimental tests need to be carried out to clarify whether the bond strength is dependent on the embedment length or not.

6. CONCLUSIONS

A series of pull tests were conducted to determine the critical embedment length of a specific type of 20 mm rebar bolt that is widely used for rock support in tunnels and underground caverns in Norway. It was found that the critical embedment length is linearly proportional to the water-cement ratio and the UCS of the grout. The critical embedment length is approximately 25 cm for the grout of UCS 37 MPa (water-cement ratio of 0.40), 32 cm for the grout of UCS 32 MPa (water-cement ratio of 0.46) and 36 cm for the grout of UCS 28 MPa (water-cement ratio of 0.50), for the Rescon zinc rock bolt cement used in the tests.

The bond strength is linearly proportional to the UCS of the grout. The bond strength of the rock bolt, that is, the average shear stress at the bolt-grout interface when the bolt starts to slip along the interface, does not seem to be a constant but increases with the embedment length as long as the yield strain of the bolt shank is not significantly large, for instance <10%. More experimental tests need to be carried out in the future to clarify whether the bond strength is dependent on the embedment length or not.

ACKNOWLEDGEMENTS

The authors would like to thank Mr Gunnar Vistnes and Mr Torkjell Breivik for their valuable assistance in the laboratory. The support by the Norwegian Road Authority–Statens vegvesen is gratefully acknowledged.

REFERENCES

1. B. Benmokrane, A. Chennouf, G. Ballivy **Study of bond strength behaviour of steel cables and bars anchored with different cement grouts** Kaiser, MaCreath (Eds.), Proceeding of Symposium on Rock Support in Mining and Underground Construction, Balkema, Rotterdam (1992), pp. 293–301

2. B. Benmokrane, H. Xu, E. Bellavance **Bond strength of cement ground glass fibre reinforced plastic (GFRP) anchor bolts** J. Rock Mech. Min. Sci. & Geomech. Abstr., 33 (5) (1996), pp. 455–465

3. J. Bjurholt **Pull and shear tests of rockbolts** Master's thesis 2007:208 CIV Luleå University of Technology. (2007) 72p. (in Swedish)

4. Y. Chen, C.C. Li **Performance of fully encapsulated rebar rockbolts and D-Rockbolts under combined pull-and-shear loading** Tunn. Undergr. Space Technol., 45 (2015), pp. 99–106

5. J.A. Franklin, P.F. Woodfield **Comparison of polyester resin and a mechanical rockbolt anchor** Trans. Instn. Min. Metall. (Section A: Min. Indust.), 80 (1971), pp. A91–A100

6. J.A. Franklin, *et al.* **ISRM suggested methods for rockbolt testing. Document No. 2 of the committee on field tests (Final draft)** E.T. Brown (Ed.), Rock Characterization, Testing and Monitoring – ISRM Suggested Methods, Pergamon Press (1974), pp. 161–168 1981

7. A.J. Hyett, W.F. Bawden, G.R. Macsporran, M. Moosavi **A constitutive law for bond failure of fully-grouted cable rockbolts using a modified Hoek cell** J. Rock Mech. Min. Sci. & Geomech. Abstr., 32 (1) (1995), pp. 11–36

8. F. Ito, F. Nakahara, R. Kawano, S.-S. Kang, Y. Obara **Visualization of failure in a pull-out test of cable bolts using X-ray CT** Constr. Build. Mater., 15 (2001) (2001), pp. 263–270

9. Kılıc, E. Yasar, A.G. Celik **Effect of grout properties on the pullout load capacity of fully grouted rock bolt** Tunn. Undergr. Space Technol., 17 (2002), pp. 355–362

10. C.C. Li, G. Stjern, A. Myrvang **A review on the performance of conventional and energy-absorbing rockbolts** J. Rock Mech. Geotech. Eng., 6 (2014), pp. 315–327

11. G.S. Littlejohn **Rock anchorage practice in civil engineering** Kaiser, MaCreath (Eds.), Proceeding of

Symposium on Rock Support in Mining and Underground Construction, Balkema, Rotterdam (1992), pp. 257–268

12. Soni **Analysis of Swellex Bolt Performance and a Standardized Rock Bolt Pull Test Datasheet and Database** Master's thesis University of Toronto, Canada (2000) 134p

13. Stillborg **Professional Users Handbook for Rock Bolting** (second ed.)Trans. Tech. Publications (1994)

14. G. Stjern **Practical Performance of Rock Bolts** Ph.D. thesis University of Trondheim, Norway (1995)

15. Y. Zhao, M. Yang **Pull-out behavior of an imperfectly bonded anchor system** Int. J. Rock Mech. Min. Sci., 48 (2011), pp. 469–475

5
CHAPTER

Components of a Wind Tunnel Balance: Design and Calibration

Miguel A. González[1], José Miguel Ezquerro[1], Victoria Lapuerta[1], Ana Laverón[1] and Jacobo Rodríguez[1]

[1] Escuela Técnica Superior de Ingenieros AeronáuticosUniversidad Politécnica de Madrid, Spain

1. INTRODUCTION

The aim of wind tunnel tests is the simulation of the flow around bodies or their scaled models. In aeronautical applications, the measurement of aerodynamic loads in a wind tunnel, forces and momentums, is a very difficult task due to the required accuracy. The wind tunnel balances, comprised by several hardware and software components, provides directly the pursued measurements, with high accuracy and reliability. For these reasons, among others, wind tunnel balances have become a common tool in testing facilities.

This chapter starts with a general description of wind tunnel balances. The number of measuring components and the position of the balance with relation to the model and wind tunnel chamber determine the wind tunnel balances designs. The most flexible ones, in terms of usability, are the six components external balances, so these will be referenced for introducing the calibration process, which is one of the key points to achieve the required aerodynamic tests results accuracy and reliability. Because of its influence on the drag measurement accuracy, the coupling effect between lift and drag measurements is analysed very deeply as well. The analysis of the non-stationary effects are finally done taking into account the wind tunnel balance requirements and constraints, with special attention on an issue not commonly mentioned, the inertia forces generated on the balance by the model vibrations, and their influence on the aerodynamic forces to be measured.Several mentions to signal processing and acquisition are done, as this is the other key point on the measurements accuracy. However, it is easy to extrapolate these procedures to other types of balances, as the main intention is to show which are the critical points that make wind tunnel balances such a special and complex hardware.

We do not intend here to describe the design and calibration procedures of the industrial manufacturers. This is the result of a work done in the Universidad Politécnica de Madrid (UPM), and the Instituto Tecnológico y de Energías Renovables (ITER, Tenerife, Canary Island, Spain, http://www.iter.es). Nevertheless, we do consider that is a good guide for developers of wind tunnel balances in institutions like UPM and ITER, where research and education are very important points.

2. WIND TUNNEL BALANCE

The wind tunnels main function is to provide flow simulation on a model introduced in a fluid flow. Global forces and momentums on the model are mainly obtained by using different wind tunnel balances; although in special tests, local balances or pressure distribution measurement can be used as well. Range, accuracy and response time of the measurements are the main parameters that define such systems. The wind tunnel balances are extensively used and are an accurate method for measurements acquisition, with a wide range of measures and a fast response to loads changes. This system requires an important initial calibration effort but once the measurements are probed to be correct, the system can be used to test several low cost models with a reduced effort. Other option for aerodynamic load measurements is the pressure measurement in several model points by means of a pressure scanner or scanivalve system. This system requires a very complex and expensivetest model. The measurement points are built in the model surface by making holes and connecting them with the scanivalve by means of a tube that transfer the pressure. These holes introduce also modifications in the flow around the model thus modifying the real behaviour of the model.

There are several types of wind tunnel balances. The most important are:

- External balances: They are placed outside the model, inside or outside the wind tunnel chamber test section, but they always introduce some interference in the wind flow. However the possibility to change test models with almost no effort provides a high flexibility to the wind tunnel facility. There are several degrees of complexity for these balances, depending mainly on the number of measurement channels, which can vary between 1 and 6.
- Internal balances: They are placed inside the model, thus no interferences are introduced in the wind flow by the balance components, but a mechanical support for the model is always needed to maintain it in the test chamber and change the model orientation if desired. The complexity of the test model is comparable or higher than the models for scanivalve systems, as the balance has to be installed inside. Thus this option does not provide flexibility in testing different models. These balances are normally supplied to the customer already calibrated and with the acquisition system. The number of measured components can also vary between 1 and 6.Figure 1 shows an example of internal balance.
- Rotary balances: Used for propellers, helicopter blades and other rotating models.

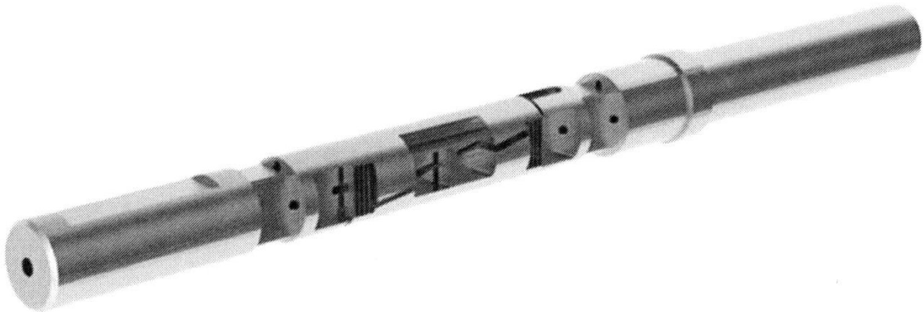

FIGURE 1. Internal balance example. Image courtesy of STARCS.

We now focus our attention in the 6 components external balances, as they provide 3 forces and 3 momentums measurement and a high flexibility for a multi test wind tunnel. Nevertheless the results presented in this report can be applied to almost all types of balances.

2.1. SIX COMPONENTS EXTERNAL BALANCE

There are several wind tunnel balances manufacturers, who have solved the problems presented in section 3 and they produce internal, external and rotary balances from the range of some Newtons up to some thousands of Newtons. But expensive commercial balances are not the only option to get an accurate wind tunnel loads measurement system. Taking into account our university environment, we, together with the ITER (www.iter.es), decided to manufacture our own balance for the ITER wind tunnel. The following chapters will show, as example to clearly explain the already mentioned key points, our experience, the problems we found during the calibration process and the solutions we have implemented, that will be beneficial for those who intent to build its own balance, as well as for those who want to understand the complexity of wind tunnel balance systems.

The commercial balances are usedto have 6 complete coupled channels and a 6x6 decoupling matrix is provided by the vendor.

This means that the forces and momentums measurements are obtained through a complex combination of the 6 channel signals by the use of a completely full decoupling matrix. The main inconvenient of this concept is that a recalibration of the balance requires a special calibration facility that is very expensive (see Figure 2).

The opposite balance design concept is a complete mechanical decoupling of the 6 loads components. Each channel output will be a unidirectional load signal (traction-compression), and the composition of the signals will be the lift, drag, lateral forces and the pitching, balance, yawing momentums. One possibility could be to use three vertical sensors to obtain the lift force (which is the dominant force in aeronautical applications), the pitching momentum and the balance momentum, two horizontal sensors in the wind flow direction to acquire the drag force and the yawing momentum and one horizontal sensor perpendicular to the flow direction to measure the lateral forces. The calibration of these balances is much easier as the matrix is almost diagonal. Nevertheless, in practice the sensors absorb lateral forces and this results in a certain coupling effect. This coupling is particularly relevant between lift and drag.

FIGURE 2. External balance example on the left. Calibration facility example on the right. Images courtesy of STARCS.

Besides the model supports, a balance support is needed to fix the balance to the wind tunnel structure. These supports are placed outside the wind tunnel test chamber, so their dimensions and weight are not important, the requirement imposed on them is an elevated stiffness.

Also important is to notice that once the external balance is providing good results for aerodynamic purposes, which are very exigent, it will provide also good results for most of other test models, as per example, cars, cyclists, buildings... so the use of this type of balance provides a multi-purpose test facility.

2.2. DATA ACQUISITION COMPONENTS

The data acquisition system (DAQ) is compounded by several items that allow the conversion of physical forces into digital values that can be managed by a computer. Several components comprise the DAQ andthe selection of their properties will have an important impact on the system behaviour. The main components are the following:

1. Mechanics: The mechanics have to ensure that no friction forces are appearing thus and such not desired loads are avoided.
2. Sensors or transducers:The sensors are devices that convert the physical load in an electrical signal. The capability of the sensors to measure the force depends on the properties of the load and the sensor itself. Several types of transducers exist (Gorlin & Slezinger, 1964):
3. Weight balanced sensors
4. Pneumatic and hydraulic sensors
5. Electromagnetic sensors
6. Spring sensors

7. Strain-gauge sensors: These are very common transducers, also called load cells, which generate the electrical signal trough the mechanical deformation that causes the applied force. Usually 4 strain gauges in a Wheatstone bridge convert the deformation into electrical signals. There are many construction types but the Z shape ones have demonstrated in our own manufactured balance,that they are sensitive and accurate, with a relatively high stiffness.

8. Electronic amplifiers: The sensor output signal is a weak signal of around a few millivolts and its power has to be increased by means of an amplifier in order to have a distinguishable signal in the analog to digital converter.

9. Wires: The electrical signal is transferred from the sensor to the amplifier and from the amplifier to the sensor by means of wires, which are desired to be of high quality and efficient.

10. Analog to digital converter (ADC): These devices convert a continuous analog signal (voltage or current) into a discrete time digital signal. Several parameters differentiate the ADC, the most important ones are the resolution, the sampling rate and the accuracy. As demonstrated in step 3.1, an ADCwith a resolution of 16 bits (2^{16} analog-to-digital conversion levels), 8 channels and a sampling speed of some 2.500 Hz per channel, result to fit with the wind tunnel balances requirements.Nowadays the hardware available at normal prices has enough quality to proportionate low error and thus good accuracy. Also the level of integration with computers is very high and manufacturers provide their own software.

11. Software: At the end of the data acquisition chain we encounter the software applications that post process the signal measurements in order to obtain the forces and momentums values. Several options are available, from commercial applications to homemade applications, from

calculus sheets to complex programming codes, which have to be selected in function of the wind tunnel complexity.

3. WIND TUNNEL BALANCE CALIBRATION

At this point, the importance of the wind tunnel balance calibration for eliminating errors in the measurement acquisition is obvious. The calibration, and recalibration if needed, is absolutely indispensable to eliminate the coupling effect and to determine the DAQ behaviour. Therefore the first step is to perform a static calibration, as explained below, and after that it is desirable to perform a dynamic calibration with the help of the already existent typical test models results.

3.1. STATIC CALIBRATION

The static calibration of the balance is not a trivial step in the chain to achieve reasonable system accuracy, although it could seem so in a first sight. As static suggest, this calibration can be done with a standalone balance, without the presence of the wind tunnel itself, and this is the standard procedure for industrial balance manufacturers. Making a calibration with the balance dismounted from the wind tunnel is much easier than calibrating it attached to its mounting supports on the tunnel. The accessibility to the balance to test every axe, in every direction and in both ways, in order to identify the behaviour of the acquisition system (mechanics, load cells, amplifier, wires, DAQ, software) is definitely easier.

Nevertheless, as shown in Figure 2, the calibration facility is very sophisticated and expensive. As this chapter treats about a self-designed and constructed balance, such calibration facility is not available, so another procedure is necessary. The gained experience shows that, in this case, instead of a standalone calibration, it is better and more accurate to do it with the balance installed in the wind tunnel. This calibration philosophy ensures that the reference axes of the balance are the same that those of the test chamber. In principle it can be though that once the balance is calibrated outside of the test chamber, it can be mounted properly by determining the local vertical and horizontal directions of the test chamber, but this may be a big mistake due to the following two reasons:

a. The balance mounting actions could ensure a good alignment between balance axes and test chamber axes, but in any case some deviation could be expected. Although these deviationsare rather small, in the case of wind tunnel testing were require very high accuracy, theycould be catastrophic for the test results. To show the importance of an angular deviation, let's assume that only a 1 degree deviation has occurred between balance and test chamber vertical direction. Let's assume also that the test chamber vertical direction corresponds with the lift direction and, correspondingly, the test chamber horizontal direction corresponds with the drag direction. The drag measurement is where there is a major impact on results,for example, if a 150 N lift force is applied to the balance, around 2,6 N are incorrectly measured in the drag direction due to the misalignment. In a model of a medium speed aircraft, the drag corresponding to a 150 N lift would range from 10 to 12 N, meaning that the measured drag value has a deviation of roughly 25 %, which is an unacceptable error for aerodynamic model tests.

b. The wind flow direction inside the test chamber has to be determined. We cannot assume that the wind direction and test chamber horizontal direction are the same, as the risk of being committing a same ordererror as the presented in the previous step is not acceptable.

A correct wind tunnel balance calibration must be preceded by some steps that allow the determination of fixed and reliable axes. The first action has to be the installation of the balance in its fixed final position in the test chamber. The second action has to be an accurate determination of the local wind flow direction inside the test chamber.

The flow direction determination can be done with the help of different instruments; the most common are the cobra, the wedge, the five holes and the cylindrical probes. All of these probes based their detection method on the same concept. The probes are a bundle of tubes where the leading edge of one of them is perpendicular to the flow direction while the leading edge of the others, one or two pairs, are cut at a known angle to the flow direction. The probe is placed in the test chamber, making an angle with the flow direction, then the pressure measured in the external probes (see Figure 3) are different, and through a precalibrated curve or by reorienting the probe, the angle of the flow can be determined.

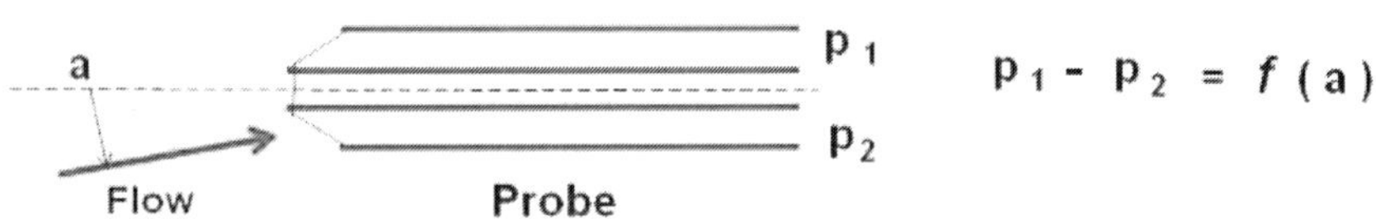

FIGURE 3. Vertical cut of three tubes showing the leading edges and the angle determination concept

Both attack angle and yaw angle can be determined by using one probe (with two pair of external tubes) or two probes (with one pair of external tubes) (see Figure 4). The leading edge angle to the flow direction determines the sensitivity of the probes to the changes in the flow angle, the dynamic pressure or the Reynolds number (Glenn LTP NASA and University of Cambridge web pages).

Once the flow direction is fully determined, the lift and drag forces directions are now known. The lift force perpendicular and the drag force parallel to the wind flow. The axes in which the forces (and momentums) have to be measured are properly defined.

Once checked that the flow is horizontal, the system is now ready for the calibration process. As aforementioned, we do not have specific calibration facilities. To avoid this problem, we have implemented some special calibration procedures, which are time consuming, but they give very accurate results.

As the function of the balance will be the measurements of aerodynamic forces, the calibration method should introduce loads in the proper direction and sense. Thus, the loads are introduced in the lift direction and not in weight direction, as this is the most critical direction while working with low drag aerodynamic models (for other applications the load direction will be of no issue). In order to ensure that we are loading the balance exactly in the appropriate direction, a specific system has been design.

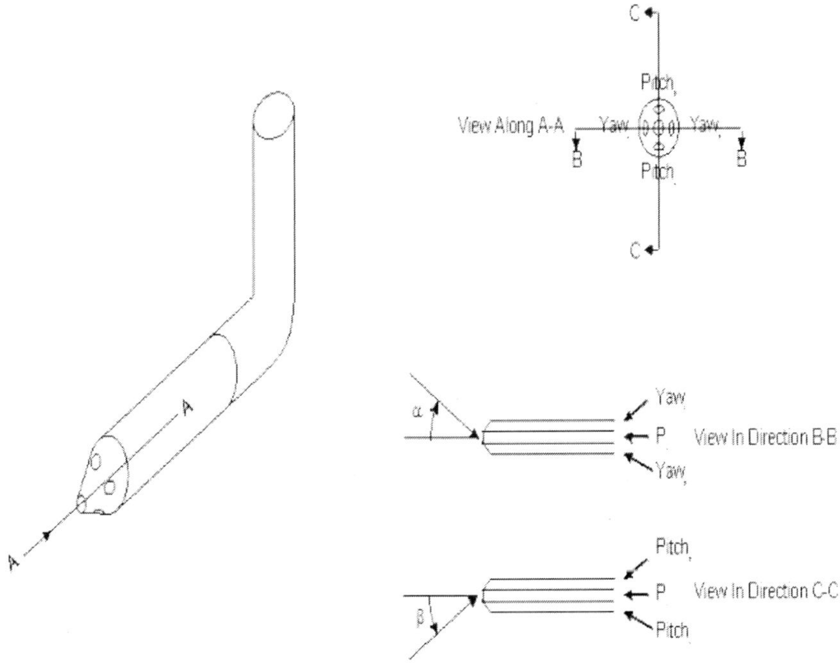

FIGURE 4. A five hole truncated pyramid probe

The procedure to be shown here takes into account, that we use standard and calibrated load cells to measure the forces on each bar. Nevertheless, we apply known forces in the direction of the three axes to check not only the calibration of the load cells, but also the signal amplifier and the data acquisition system. As we expected, the manufacturer cell calibration and our calibration had a very good agreement, differences were less than 0,1%. It probes as well, that our philosophy of mechanical decoupling is a very good approach for the design of external balances.

It is important to point out that, at this moment, we were interested only in one direction each time. So small errors in the vertical or both horizontal directions (wind direction and transverse) did not introduce important errors on the true loads, because the cosines of the expected deviation angles are practically 1,0.

As abovementioned, for the current application, the only relevant coupling is between drag and lift. Taking into account that the flow in the test section is horizontal, an special device was designed to ensure that we apply a vertical load to the balance, with an error less than 0,1 mm in 600 mm. In these conditions, the coupling can be measured and corrected, and for a lift coefficient of 0,5, which is normal for cruise conditions of low speed aircrafts, the effect on drag is in the order of one drag count.

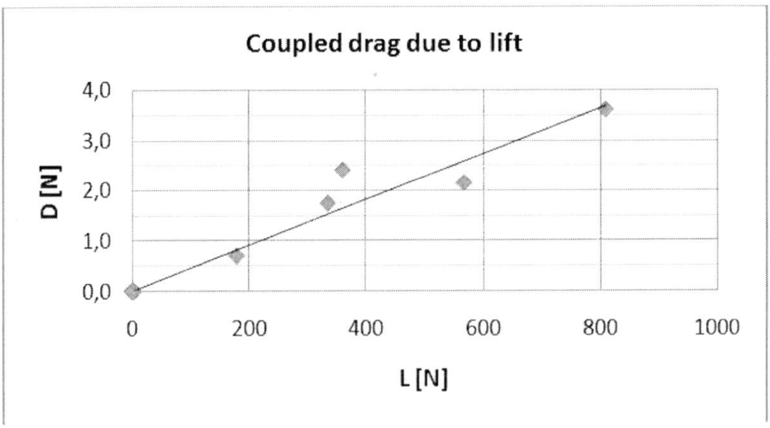

FIGURE 5. Coupled drag due to lift. Horizontal axis represents the applied vertical force, and vertical axis the measured horizontal component due to coupling effect.

Figure 5 shows the results of the drag due to lift coupling tests, the *x* axis represents the loads in the vertical direction (lift) and the *y* axis the measured in horizontal direction (drag). It shows that the coupling effect is almost linear, being the drag 0,46% of the applied lift.

3.2. DYNAMIC CALIBRATION

Finally, to check all the system working in real conditions, we could use a reference model, such as a two dimensional model of the NACA 0012 wing section, to make final adjustment of the DAQ and coupling corrections. A well known standard calibration model would be another option, but the L/D ratio is well below the range we expect to make our tests. In both cases, differences in turbulence level and/or surface smoothness may produce differences of the same order than those to be checked.

Due to those reasons and the cost, in terms of time and money, it was finally decided to make a deep analysis of the measured results, to determine the DAQ requirements needed to get a high confidence level on the measured forces. The complete procedure is presented in the following paragraph.

4. WIND TUNNEL BALANCE REQUIREMENTS

The better way to understand the strong requirements demanded to a wind tunnel balance is by studying the following example, where rough numbers involved in the problem are estimated.

In a 2,0x2,0 m low speed wind tunnel test section, with a testing velocity of around 50 m/s, using a model of a medium speed aircraft at cruise conditions, the lift would be roughly 150 N and the drag would range from 10 N to 12 N, for the high aerodynamic efficiency condition. For this type of model aircraft, the accuracy of the drag coefficient must be greater than 0,5 drag counts, which means that the accuracy of the balance, in the drag direction, must be higher than 0,015 N.

For the case of static loads measurements, where there is only one component, accuracy may or may not be a very tight requirement. But in the case of the wind tunnels, where we intend to measure simultaneously up to 6 components, forces and momentums, on a model under aerodynamic loads, with some non-stationary terms due to vibrations; amongst others, the following problems can be expected:

a. The model vibrations induces inertia forces on the balance.

b. There are aerodynamic forces on the model support system and interferences between these and the model may appear.

c. The complete mechanical decoupling of the forces components is very difficult and expensive; therefore some coupling between lift and drag may appear.

During the setting of a new self-designed balance, all these points have been deeply treated, and the acquired experience is the base for this paragraph.

4.1. INERTIA LOADS ON THE BALANCE

When testing an aerodynamic model in a wind tunnel, the forces and momentums measured come from different sources. The most important ones are the aerodynamic forces and momentums themselves, but some other undesirable loads could be expected. Although wind tunnel balances are necessarily very stiff, due to the model support and other elements, it is difficult to avoid oscillatory movements of the model, due to the aerodynamic loads. Those vibrations induce inertia forces and momentums on the balance, and might have a negative impact in the validity and/or accuracy of the measurements. Also some intrinsic data acquisition systemnoise has to be expected. Inertia forces and electronic noise are not desired, but measured, together with the aerodynamic loads; as those are not present in a free flight model, the study and evaluation of their impact on the pursued results is very important.

The inertia loads that appear in wind tunnel tests will depend strongly on the test facility and its components, so a specific study of each configuration should be performed. Nonetheless, the procedure, the critic analyses of the measurements and the influence range of these not desired loads are similar for all the wind tunnel tests configurations, so the results presented in these paragraphs will be of high interest for wind tunnel test engineers.

We can start making some rough calculations to show the true magnitude of the inertia loads. If we consider an amplitude in the order of 0,5 millimetres and an oscillatory frequency in the flow (drag) direction of some 5 Hz, which are very representative true values, the oscillation of the model produces accelerations in the range of 0,49 m/s^2. If the mass of the model is, i.e., 50 kg, the appearing inertia forces are of the order of 24 N, which is of the same order of magnitude as the value of the drag that is to be measured. This is the reason why these inertia loads have to be carefully studied, in order to determine their magnitude, shape and relation with the aerodynamic loads.

First point to be determined is the sampling rate of the electronic signal, which is necessary for an accurate determination of the inertia terms, and thus the determination of the true aerodynamic signal. The aliasing, which is an effect that cause different sampled signals to become indistinguishable, is taken as the reference for the sample rate selection in several publications. The sampling theorem, frequently called Nyquist sampling theorem, states that a continuous signal can be properly sampled if the following expression accomplished (Smith, 1997):

Sampling Frequency2*Band WideSampling Frequency2*Band Wide (1)

Following this rule, if we have a frequency signal of around 5 Hz, it can be stated that a sample rate of more than 10 Hz is enough for our purposes. As we will show, this is not enough in our case, due to several reasons, between them: the signal frequency study will not be accurate and the sample rate will impact the measurement results. The sampling rate should be high enough to get accurate results and avoid the previous issues. Nowadays, the sampling frequcncy of the wind tunnel balance systems is much higher than the maximum expected frequency of the signal induced on the balance by the aerodynamic and inertia forces, thus those mentioned problems are overpassed, as well as the aliasing effect. Also with today´s computer capabilities, managing an elevated quantity of data is no longer a problem. High sampling frequency produces benefits in reducing the random noise originated in the acquisition system (DAQ) as well.

While using low sampling rates, it is important to ensure that the sampled signal comprises complete cycles, so that no partial cycles are included in the signal mean and standard deviation calculation. The values obtained if, as an example, a half wave is included, could cause significant deviations with relation to the correct results. However, this effect is only recognizable using low sapling rates, and disappears if a high sampling rate is used.

In order to ensure that the noise of the measured signal is low enough for our purpose, the signal to noise ratio (SNR) has to be higher than 3, condition that ensures that every signal is recognizable. Although this is a soft requirement to the hardware currently available, there is always present a noise signal that is a not desired random signal. One simple and efficient method that can be applied to cancel out the noise signal is the time-averaging. This method, as presented in several publications (Lohninger, H., 2010), consist on registering and summing up the signal repeatedly, an action that applied to a stable signaleliminates the noise signal (our aerodynamic loads and inertia loads are supposed to be stable over hundreds of cycles once the system reaches a steady state). A high sample rate impacts positively in this noise reduction method.

Before starting aerodynamic tests, some mechanical tests were done to determine the main frequency of the complete system, balance plus model. Two free vibration movements were induced, in the flow direction and in yaw oscillation. In both conditions, the reactions on the balance were measured. The signal analyses showed that the main frequency for such unforced movement was, roughly, 5,2 Hz. In principle, this is the main frequency to be considered for the following analysis, and it will be shown that this main frequency is not modified by forced oscillations, due to aerodynamic loads.

The determination of the minimum sampling rate that provides accurate results has to take into account several considerations. The most important ones are, as already mentioned, the accuracy of the results and the main inertia frequency determination. The first measurements acquisitions, although the main inertia forces frequency is already known, have to be taken with a high sampling rate, at a minimum of 2 orders of magnitude above the expected vibration frequency (even 3 orders of magnitude are recommended for the complete analysis). As explained later on in this text, this will allow a proper frequency domain study, as no frequency components are lost due to a low sampling rate. The focus of attention is the accuracy of the results. The sampling rate has an important impact in the mean of the measured signal, which, at the end of the process, is the aerodynamic load we want to measure. Through the following example the influence of the sampling rate is clearly shown. Figure 6 shows the mean value of a real wind tunnel signal calculated each second for total signal duration of 20 seconds (blue line). The signal is the same for all the plots and the only difference is the sampling rate, from 2.500 Hz, near 3 orders of magnitude above the mechanical main frequency, to 10 Hz, double of the main frequency. The linear regression of those average values is also plotted in all the figures. The last plot is the representation of the whole average signal mean value, for the complete 20 seconds, versus the sampling rate.

As shown in Figure 6, for low sampling rates, 10 Hz and 20 Hz, which have the same order of magnitude than the Nyquist sampling rate for this signal, the accuracy and reliability of the mean value is poor. The linear regression and the mean value have important deviations. However, a sampling rate of 100 Hz gives good results although this frequency, as shown in the plot of the averaged force mean value, is at the limit. For higher sampling rates as 500 Hz and 2500 Hz the accuracy of the force mean is very good.

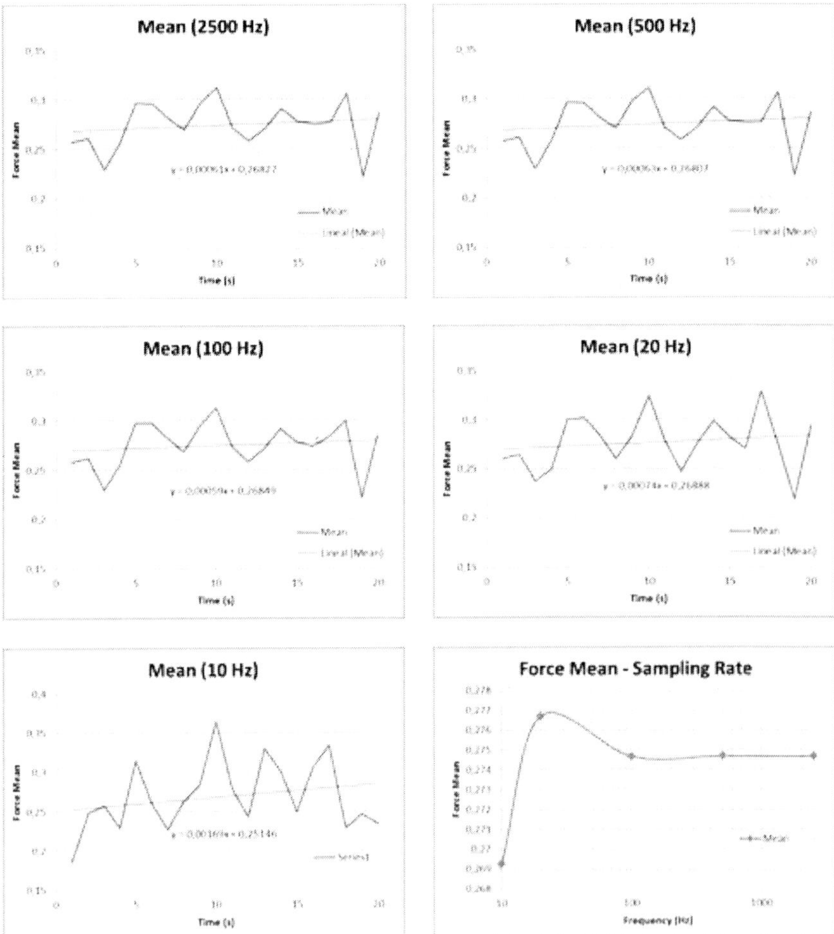

FIGURE 6. Plots of a signal force mean calculated each second. Five plots for different signal sampling rates: 2.500, 500, 100, 20 and 10 Hz. Plot of the whole averaged force mean value versus sampling frequency.

Then, it can be stated that the minimum sampling rate to get accurate results has to be at least 2 orders of magnitude above the Nyquist sampling rate. Even if the sampling rate used were very high, it is always recommended to perform this study, to understand the complete phenomena that occurs.

Once the appropriate measurements are acquired, the signal properties can be studied. It is interesting to know the frequency components of a signal (in example the inertia forcesfrequency) but their identification by its representation in the time domain usually is a difficult task. A deeper study of the measured signals trough the Fourier transform provides the frequency components of the signal. One fast approach to this Fourier transform study is to perform a discrete Fouriertransform, which is identical to samples of the Fourier transform at equally spaced frequencies, by using the fast Fourier transform method (FFT) (Oppenheim & Schafer, 1989). In the frequency domain, it is easy to identify each of the different signal sources. With the aim to clearly show the influence of each measurements source, real measurements will be presented below. Those were taken at the ITER wind tunnel, that is provided with an own manufactured balance, which is an accurate wind tunnel 6 mechanically decoupled components balance.

ITER Low Speed Wind Tunnel (ITER-LSWT) has a 2,0x2,0 m test section, 3,0 m long. In the current configuration, which includes devices to reduce turbulence, the maximum allowable speed is 50 m/s, with a turbulence level below 0,5%. The model used for the tests is a section of a wing, still under design, to be used in a solar powered aircraft, with 667 mm cord and 1.990 mm span.

It is important to look at the signals frequency spectrum from a critical point of view. It is not enough to identify the main frequency but the amplitude spectrum of the frequencies, and their importance in comparison with the constant signal (~0 Hz frequency signal) that represents the aerodynamic force acting on the model. This is clearly shown in Figure 7, were the signal is plotted in both: time and frequency spectrum.

The main frequency of the inertia forces signal is slightly higher than 5,0 Hz, the same than for the mechanical free vibration, and its amplitude value in comparison with the complete signal amplitude is near 75%.

Most of the balances obtain the forces and momentums values from a combination of several load cells measures.Then, it can be thought that the study of the resultant load could be enough to determine the inertial behaviour of the system, but this is not true. It is important to study the frequency behaviour of each measurement channel signal, before studying the combination of the measurement channels.Figure 8 shows the resultant signal of the two cells that measure the drag component in the ITER-LSWT balance. In this case, although there is a pick of frequency for the same value than the single analysis, it should be stated that the main inertia movement is occurring at around 10 Hz.

It is clearly shown that in the combined signal, although a small peak appears at 5,0 Hz, the main vibration frequency is 10 Hz. However, Figures 9 and 10 show the 2 measured signals whose combination result in the previous drag force signal.

This happens because of several coincidences: the model is symmetric, both load cells are placed equidistant along the symmetric model and the inertial vibration axe induced on the model by the aerodynamic flux is perpendicular to the load cells plane. Thus, when both channel signals are summed up, the inertia components are cancelled. This is a very good example to emphasize the necessity to study signal channels independently, prior to forces and momentums calculation, although the extreme case, with other model or in a different wind tunnel facilities configuration this behaviour could not happened.

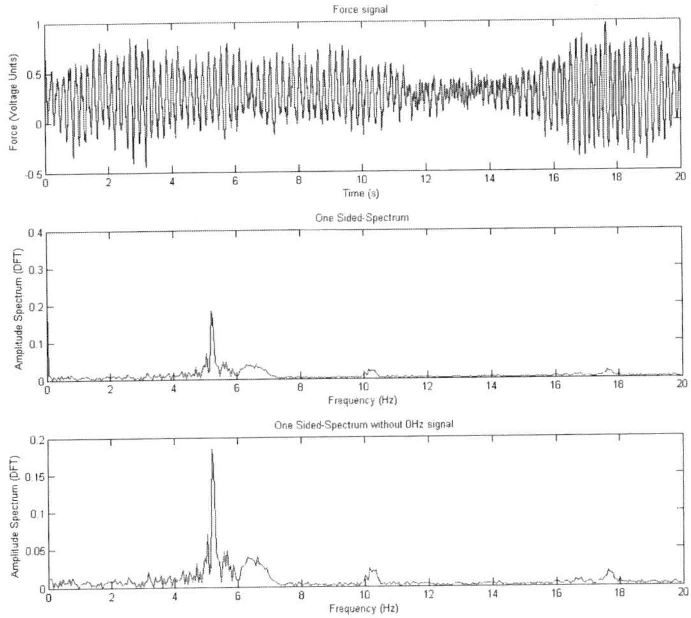

FIGURE 7. Representation of a measured signal, in voltage, at 2.500 Hz sampling rate of an aerodynamic model in the ITER-LSWT, at a wind speed of 20 m/s and at angle of attach of 8º. Time domain signal plot during 20 sec. Frequency domain signal plots with different amplitude scales, with and without the 0 Hz component of the signal.

Due to the particular characteristics of aerodynamics models, the influence in the drag measurements is much more important than in the other forces and momentums, such as the lift measurements, so it is highly recommended to focus the attention in those channels related to the drag force measures.

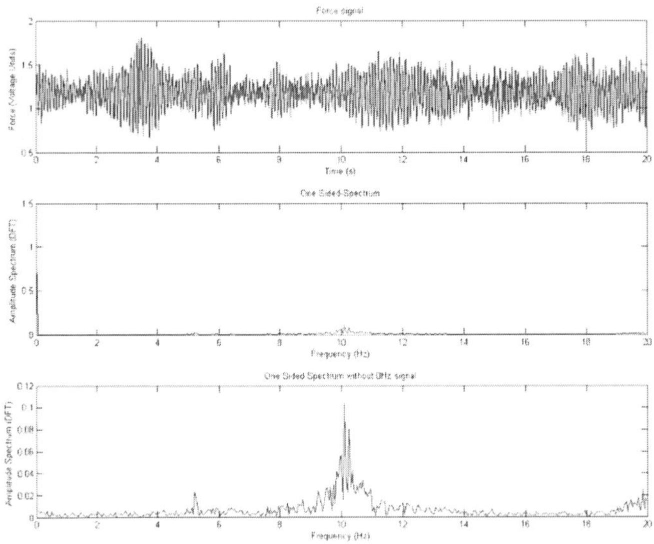

FIGURE 8. Representation of a resultant force signal, combination of two drag cells, at a 2.500 Hz sampling rate of an aerodynamic model introduce in the ITER-LSWT, at a wind speed of 40 m/s and at angle of attach of 8º. Time domain signal plot during 20 sec. Frequency domain signal plots with and without the 0 Hz component of the signal.

4.2. MODEL SUPPORT INTERFERENCES

The external balances are used as fix element of the wind tunnels, however the test models are interchangeable items. The model support system is the mechanical link between the balance and model. There are several types and sizes of supports,but all of them may interfere with the aerodynamic measurements. Two origins of interferences can be highlighted:

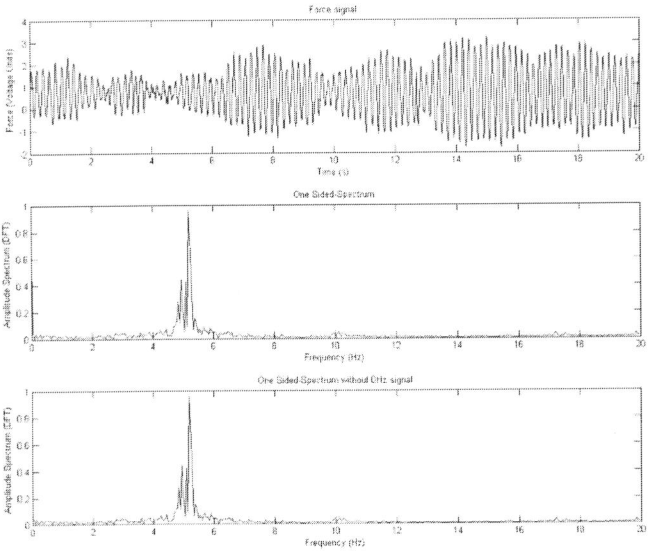

FIGURE 9. Representation of channel 1 signal at a 2.500 Hz sampling rate of an aerodynamic model introduce in the ITER-LSWT, at a wind speed of 40 m/s and at angle of attach of 8º. Time domain signal plot during 20 sec. Frequency domain signal plots with and without the 0 Hz component of the signal.

a. Aerodynamic forces on the model support system.

b. Aerodynamic interference between the model support system and the model itself.

The occurrence and impact of each of the interferences depend on the shape, size and disposition with relation to the model, but in almost all cases, mainly in aeronautical purposes, the model support system is another source of problems for an accurate drag measurement.

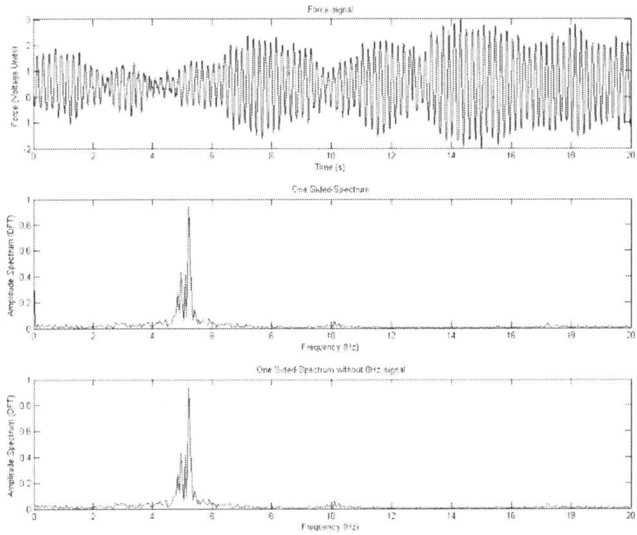

FIGURE 10. Representation of channel 2 signal at a 2.500 Hz sampling rate of an aerodynamic model introduce in the ITER-LSWT, at a wind speed of 40 m/s and at angle of attach of 8º. Time domain signal plot during 20 sec. Frequency domain signal plots with and without the 0 Hz component of the signal.

A direct model fixation to the balance might be the best option in order to avoid interferences, but this solution will work in few and very specific cases, such as, half wings or half model aircraft. In all other cases of interest, as wings, nacelles, complete aircrafts... tests, the presence of a more or less complex support cannot be avoided; as the model has to be placed more or less in middle of the test chamber, and the aerodynamic forces have to be transmitted to the balance. Moreover, it is very common that for aeronautical tests, as a minimum, the angle of attack of the model should be variable, which introduce more complexity to the support and, therefore, more interferences.

Excluding a few cases, a support system is unavoidable. Its influence quantification and the methods to subtract the interferences are key points to gain measurement accuracy. If a very stiff support is used, its drag and interferences on the model may be very important, once more in the order of the values to be measured for cruise conditions. On the other hand, the size of the support fairings will be smaller if we just faire the supports, but in this case the supports drag becomes very important too.

There are several methods to subtract the contribution of the support to the balance forces and momentums measurements. Three methods with a complete different concept are presented here:

- Experimental method: By taking series of measurements before placing the model, it is possible to determine the support aerodynamic contribution. However, by means of this method, the possible aerodynamic interferences between support and model are not contemplated. In case the supports wake does not impact on the model, and vice versa, the mutual influence would be negligible, but in other case, the support-model interference might be studied with more detail. This is a very common method, as other none desired loads are eliminated as well, from the model and support weight.
- Theoretical method: Theoretical calculations of aerodynamic forces and momentums over the supports can be subtracted to the wind tunnel measurements. The supports fairing geometry should be known and common, with an extensive and contrasted bibliography, so that the aerodynamic forces and momentums theoretic calculus will be reliable enough. The support fairings airfoil should be a symmetric one, and its angle of attack should be 0º in order to generate only drag force and one momentum component. Also, the possible interferences between support and model and vice versa have to be assessed.

- Hybrid method: It is a mix of both previous methods, providing an easier experimental part and a more reliable theoretical part (Horsten & Veldhuis, 2009).

Once support forces and momentums has been measured or computed, it is convenient to obtain the aerodynamic forces and momentums coefficients, made non-dimensional using the same characteristic surface and dimensions than for the model, in order to be able to substrate those coefficients to the measured ones. It is also convenient to test the supports for the same Reynolds number (Re) range than the model to be tested, then it should be possible to obtain a fitted curve for each coefficient, normally depending upon Log(Re).

4.3. COUPLING EFFECTS

The balance that has been used in the ITER-LSWT facilities was designed with the philosophy of mechanical decoupling of forces. So that, the measurement of loads on six traction-compression bars allows obtaining all forces and momentums components. From the theoretical point of view, this design method supposes that there is no coupling in the three forces components, because bars cannot support any transverse load. Nevertheless, it is not absolutely true, and some coupling effects among measured loads must be expected. There are several sources that could contribute to the coupling effect. The most relevant are:

a. Deformations of the balance forces transmission rods.

b. Deformation of the model supports.

c. Inaccurate assembly of balance components.

These errors can be eliminated by means of a correct calibration, as presented in step 4, but it is important to know their magnitude in order to stress the importance of this calibration.

This coupling effect is very critical in the relation between lift and drag. In a high quality balance such coupling is only about 1%,which couldseem to benegligible, but, in the case of aeronautical applications, this is the order of the ratio between lift and drag, so that such coupling must be accurately measured during the calibration process. If we consider the example introduced at the beginning of this section, a lift force of around 150 N, may produce a force of 1,5 N in the drag direction, that is in the same order of magnitude than the drag forces appearing due to aerodynamic loads.

5. CONCLUSIONS

The first tests done with our self-designed and manufactured balance has been on a two dimensional profile. Such profile will be used in a solar powered aircraft, with critical design conditions for Re=500.000 and lift coefficient, $c\,l$=0,8 and with constant cambered upper surface, in order to facilitate the installation of the solar cells. Profile aerodynamic characteristics were computed using several codes previous to its manufacturing and testing, because a minumum aerodynamic efficiency of 80 is required in the abovementioned design conditions. Experimental results have shown a complete agreement with the expected ones, including lift curve slope and aerodynamic efficiency, proving that both, balance capabilities and calibration procedures, are good enough for our requirements: a maximum error of one drag count for critical design conditions. Of course, many balances achieve errors even one order of magnitude less, but their cost is several orders of magnitude higher.

From the technical point of view, we have learned many things, but we want to emphasize only three important aspects:

- Mechanical decoupling is a very efficient design philosophy to reduce sources of errors and to facilitate balance calibration.
- Self-designed and manufactured wind tunnel external balances should be calibrated mounted in the definitive place.

- Detailed measurement and analysis of the measured signal is highly recommended to fix DAQ performances, and can efficiently replace the dynamic calibration.

6. ACKNOWLEDGEMENTS

The whole activities have taken about ten years, with the participation of many professionals and students. First of all, authors wish to acknowledge to ITER and its managing director, Manuel Cendagorta, the support given in balance manufacturing, setting and tests. In particular, we must mention to Esther Friend, because of her personal compromise and enthusiast help; the support given by Guillermo Galván, Jesús Fernández, Jesús Rodríguez, Juan Enrique Martín, and many others.

Many thanks as well to Lidia Esteban and Alvaro Lozano, students of the UPM, which have collaborate in different times of the project. The latest, but not the less important, students involved in the project have been Artur Jarzabek and Christopher Field, from the Imperial College; thanks in deed for your help in final tests performance and proof reading.

REFERENCES

1. S. M. Gorlin, I. I. Slezinger, 1966 Wind Tunnels and Their Instrumentation, Israel Program for Scientific Translations, 1680 Israel, Jerusalem

2. B. J. C. Horsten, L. L. M. Veldhuis, 2009 A New Hybrid Method to Correct for Wind Tunnel Wall- and Support Interference On-line, World Academy of Science, Engineering and Technology 58

3. H. Lohninger, 2010 Fundamentals of Statistics, Ihr Epina eBook-Team, Retrieved from <http://www.statistics4u.com/fundstat_eng/>

4. A. V. Oppenheim, R. W. Schafer, 1989 Discrete-Time signal Processing, Prentice-Hall Inc., 0-13216-776-9 New Jersey

5. S. W. Smith, 1997 The Scientist and Engineer's Guide to Digital Signal Processing, California Technical Publishing, Retrieved from <http://www.dspguide.com/>

6. Unknown author. 2010 Pressure Probes, Department of Engineering, University of Cambridge, Retrieved fromhttp://www-diva.eng.cam.ac.uk/whittle/pressure. html

7. Various authors. 2010 Beginner's Guide toWind Tunnels, Glenn Learning Technologies Project ; NASA, Retrieved from <http://www.grc.nasa.gov/WWW/k-12/airplane/bgt.html>

6

CHAPTER

Interaction between Surface Structures and Tunnelling in Sand: Centrifuge and Computational Modelling

Giorgia Giardina, Matthew J. DeJong, Robert J. Mair

Department of Engineering, University of Cambridge, Trumpington Street, CB2 1PZ Cambridge, UK

HIGHLIGHTS

- The interaction between tunnelling in sand and surface structures was investigated.
- •A 2D finite element model was validated through comparison with centrifuge results.
- A no-tension interface was needed to simulate the observed soil–structure gap.
- •A sensitivity study shows the effect of building weight on the soil–structure interaction.

ABSTRACT

Tunnelling in urban areas requires a careful estimation of the consequence of soil settlements on existing buildings. In this paper the interaction between the excavation of a tunnel in sand and surface structures is investigated. A two dimensional finite element model is presented and validated through comparison with centrifuge test results, both with and without structures. The model is then used to perform a sensitivity study on the effect of building weight on soil movements and structural deformations. The results of the validation indicate that assuming a no-tension interface between the soil and the structure is essential to capture the soil–structure interaction that was experimentally observed. The parametric analyses show that the relation between the building stiffness and the tunnelling-induced deformations depends on the building weight.

Keywords: Centrifuge testing; Numerical modelling; Settlements; Soil–structure interaction; Tunnelling

1. INTRODUCTION

Urban growth is continuously increasing and is expected to reach 60% in 2030 and 70% in 2050 (World Urbanization Prospects, 2014). One of the major challenges for expanding cities is the development of efficient and sustainable mobility. Rail and metro tunnels are an extensively adopted solution to address road congestion while minimising pollution. However, underground excavation causes soil settlement, which may affect surface structures. Tunnelling designs therefore require extensive assessment of potential settlement damage to existing buildings.

Standard procedures for the prediction of tunnelling-induced damage to surface masonry structures have been developed and applied (Mair et al., 1996). The core stage of these procedures assumes that predicted soil surface greenfield settlements caused by the excavation can be directly applied to a weightless and elastic beam that serves as a representative model of the building (Burland and Wroth, 1974, Burland et al., 1977,Boscardin and Cording, 1989 and Burland, 1995). Total tensile strains of the equivalent beam are analytically derived and compared to critical values that determine the level of expected damage. The method has been recently extended to the third dimension (Namazi and Mohamad, 2013).

The original formulation of the critical tensile strain approach neglects soil–structure interaction. For this reason, a primary development of this method was the inclusion of the relative stiffness between the structure and the soil (Potts and Addenbrooke, 1997,Son and Cording, 2005, Franzius et al., 2006, Dimmock and Mair, 2008, Farrell, 2010 and Goh and Mair, 2011). Other aspects have also be shown to affect the soil–structure interaction, including: the nonlinear behaviour of the building material (Boonpichetvong and Rots, 2005, Son and Cording, 2005, DeJong et al., 2008, Laefer et al., 2011, Nghiem et al., 2014, Giardina et al., 2012, Giardina et al., 2013 and Amorosi et al., 2014), the initial structural damage (BRE, 1995, Guglielmetti et al., 2008, Devriendt et al., 2013 and Clarke and Laefer, 2014), the presence of large openings (Son and Cording, 2007, Melis and Rodriguez Ortiz, 2001, Pickhaver et al., 2010 and Giardina et al., 2015b), the torsional behaviour of the building (Burd et al., 2000, Franzius et al., 2006, Giardina et al., 2010 and Losacco et al., 2014) and the building weight (Burd et al., 2000, Liu et al., 2000, Franzius et al., 2004, Rampello et al., 2012 and Giardina et al., 2015a). These aspects have been evaluated by using physical (Farrell, 2010, Laefer et al., 2011, Giardina et al., 2012 and Nghiem et al., 2014) and numerical (Potts and Addenbrooke, 1997, Burd et al., 2000, Liu et al., 2000, Son and Cording, 2005, Son and Cording, 2007, Boonpichetvong and Rots, 2005, Franzius et al., 2006, DeJong et al., 2008, Giardina et al., 2010, Pickhaver et al., 2010, Giardina et al., 2013, Amorosi et al., 2014 and Losacco et al., 2014) models.

However, fewer numerical modelling studies have been validated with experimental results (Son and Cording, 2005 and Giardina et al., 2013); in particular, computational modelling of the soil–structure interaction due to tunnelling-induced settlements has never been compared with centrifuge tests, which can reproduce the real stress conditions of the prototype due to the increased acceleration applied to scaled models. Centrifuge experiments have been recently performed on building models (metallic plates) of different weight and stiffness subject to tunnelling (Farrell, 2010). The results demonstrate that the soil–structure interaction may include the development of a gap between the structure and the soil. The occurrence and magnitude of the gap depends on the type of structure (particularly its stiffness and weight) and the amount of tunnelling induced ground settlement. These tests suggest that accurate simulation of the building response depends on the inclusion of the building weight and a no-tension soil–structure interface.

The effect of building weight has been numerically analysed (Franzius, 2003) under the assumption that for realistic values of weight and ground loss, no gap can develop between the soil and the building. The effect of the building weight was found to be negligible with respect to the effect of the relative stiffness. A no-tension interface between the structure and the soil has also been considered previously (Rots, 2000) to reduce the effect of a greenfield settlement profile directly applied to the structure; nonlinear interface elements between the soil and the structure have been used in subsequent numerical studies on building response to differential settlements (e.g.Boonpichetvong and Rots, 2005, Netzel, 2009 and Giardina et al., 2013).

The aim of this study is to investigate the key aspects of modelling the interaction between tunnelling in sand and surface structures and to evaluate the effect of building weight on this interaction. A 2D finite element model with nonlinear interface elements is used to identify important modelling aspects, like the simulation of the experimentally observed soil–structure gap. The model is evaluated through comparison with centrifuge experiments by Farrell Farrell and Mair (2011). The model is then used to perform a sensitivity study on the interaction between the building stiffness and weight, the latter of which is typically neglected in the current damage assessment procedures.

2. NUMERICAL MODEL

Fig. 1a illustrates the experimental setup described by Farrell (2010) and reproduced by the numerical model presented in this paper (Fig. 1b). In the centrifuge experiment, a tunnel excavation was simulated under plane strain conditions in a high resistant steel box containing fine dry sand. The model was subjected to an acceleration of 75 g and therefore reproduced the response of a prototype 75 times larger. The volume loss induced by the tunnelling and the consequent settlements on the soil surface were obtained by gradually withdrawing water from a rubber membrane installed around the tunnel; the membrane pressure was steadily increased during the centrifuge spin-up so that it was always equal to the total vertical pressure at the tunnel axis. Having reached 75 g water was then gradually withdrawn from the membrane, simulating tunnel volume loss. After a preliminary test performed in greenfield conditions, four aluminium beams of different stiffness and weight were added to the model surface (Fig. 2), to investigate the soil–structure interaction induced by tunnelling.

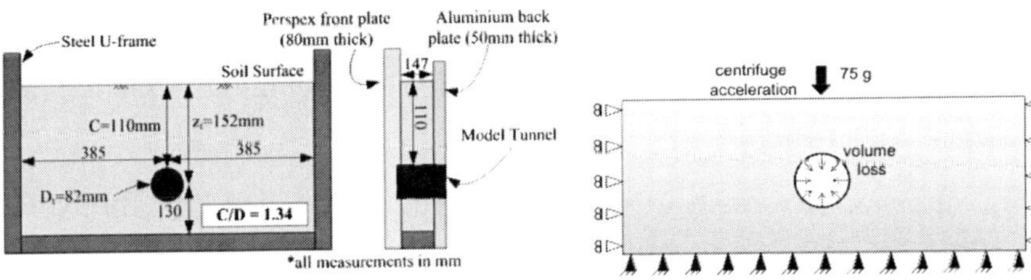

(a) experimental (Farrell, 2010) (b) numerical

Fig. 1. Greenfield model (dimensions in mm).

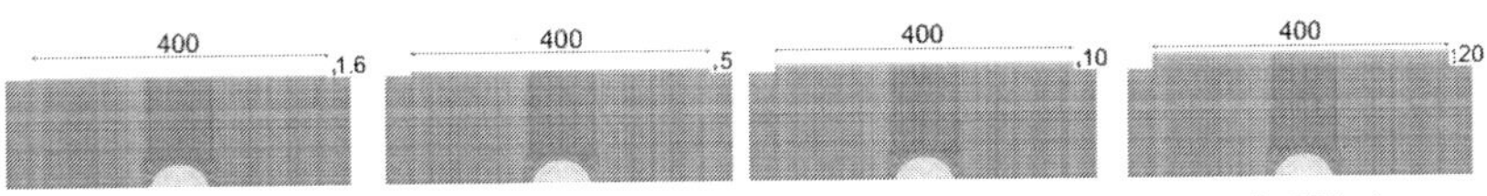

(a) STR-1 (b) STR-2 (c) STR-3 (d) STR-4

Fig. 2. Numerical model with structures (dimensions in mm).

Fig. 2. Numerical model with structures (dimensions in mm).

To reproduce the boundary conditions imposed by the strong box, the vertical free edges of the numerical model were constrained in the horizontal direction, while the bottom side was fixed in both the vertical and horizontal direction (Fig. 1b). The centrifuge spin-up was simulated by incrementally increasing the gravity load up to 75 g. During this phase the tunnel was neglected. After the gravity load of 75 g was applied, the soil elements corresponding to the tunnel section were deactivated. In order to maintain equilibrium between internal and external forces, an outward radial pressure was simultaneously applied at the tunnel boundary. The surface settlement profile was obtained by gradually reducing the pressure up to zero and then by incrementally applying an inward radial pressure. The change of pressure was proportional to the initial equilibrium value.

Fig. 2 shows the mesh of the finite element model. Quadratic plane strain elements were used for the soil, the tunnel and the structures. The aluminium beams were modelled with an isotropic linear elastic material. Quadratic interface elements with no tensile strength, high stiffness in compression and low stiffness in shear were inserted between the soil and the structure. The compressive and shear behaviour of the interface was assumed to be linear elastic. All material parameters are listed in Table 1.

Two alternative constitutive laws were considered for the soil. Firstly, a linear relationship between the Young's modulus E and the depth z was assumed:

$$E(z) = E_0 + (z - z_0)\frac{\partial E}{\partial z} \tag{1}$$

where E_0 is the Young's modulus at the soil surface, z is the vertical coordinate of the soil, z_0 is the vertical coordinate of the soil surface and $\frac{\partial E}{\partial z}$ is the gradient of the Young's modulus in the vertical direction. Secondly, the isotropic strain invariant ε_v and the shear

Table 1. Material parameters for the numerical model.

Aluminium	Young's modulus	$E_a = 70 \times 10^3$ N/mm^2
	Density	$\rho_a = 2.7 \times 10^{-6}$ kg/mm^3
	Poisson's ratio	$\nu_a = 0.3$
Interface	Normal stiffness	$k_n = 100$ N/mm^3
	Tangent stiffness	$k_t = 0.1$ N/mm^3
	Tensile strength	$f_{t,b} = 0$ N/mm^2
Soil	Reference Young's modulus	$E_0 = 25$ N/mm^2
	Young's modulus gradient	$\frac{\partial E}{\partial z} = -4.5 \times 10^{-2}$ N/mm^3
	Density	$\rho_m = 1.59 \times 10^{-6}$ kg/mm^3
	Poisson's ratio	$\nu_m = 0.25$
	Reference shear modulus	$G_1 = 1$ N/mm^2
	Reference compression modulus	$K_1 = 2.5$ N/mm^2
	Power constant	$n = 0.53$
	Reference pressure	$p_0 = -1 \times 10^{-3}$ N/mm^2

strain invariant ε_s were assumed to be dependent on the total mean normal stress p and the deviatoric stress q according to the following power-law relationship for undrained conditions:

$$\begin{cases} \varepsilon_v = \frac{1}{K_1}\left(\frac{p}{p_0}\right)^{n-1}\left(1 - \beta\frac{q^2}{p^2}\right)p \\ \varepsilon_s = \frac{1}{3G_1}\left(\frac{p}{p_0}\right)^{n-1} q \end{cases} \quad (2)$$

where K_1 is the reference compression modulus, G_1 is the reference shear modulus, n is a constant coefficient, p_0 is the reference pressure, equal to −1 kPa, and $\beta = \frac{K_1(1-n)}{6G_1}$.. The values of K_1, G_1 and n are experimentally calibrated, as noted below. The Young's modulus is derived as $E=2(1+\nu)G_1$, where $\nu = \frac{3K_1-2G_1}{6K_1+2G_1}$ is the Poisson's ratio. The adopted soil models are evaluated based on their ability to reproduce the experimentally observed tunnelling-induced settlement profile at the surface level.

Fig. 3a shows the settlement of the soil above the tunnel centreline measured at the surface (z=0) and at 60 mm below the surface (z=60) during the spin-up of the centrifuge model from 0 to 75 g. These results are used to calibrate the values of the Young's modulus gradient $\frac{\partial E}{\partial z}$ of the linear elastic soil model and the values of n, K_1 and G_1 of the nonlinear elastic soil models. The comparison with the numerical results (Fig. 3b) indicates that the nonlinear elasticity provides a considerably better fit to the experimental curves, being capable of capturing the increase of soil stiffness with the increasing confining stress.

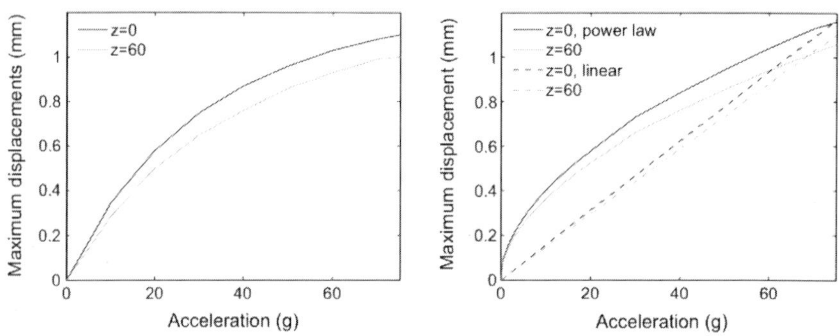

(a) experimental (Farrell, 2010) (b) numerical

Fig. 3. Maximum settlement above the tunnel during centrifuge spin-up.

3. SIMULATION OF CENTRIFUGE TESTS

In this section, the experimental results by Farrell (2010) are used to evaluate the numerical models. In particular, the comparison between experimental and numerical results is used to select the more adequate soil model, evaluate the need for using a no-tension interface between the soil and the structure and assess the model's ability to simulate the structural response.

3.1. Greenfield tests

Fig. 4 illustrates the results of the analysis in terms of surface settlements of the soil for increasing values of volume loss V_L;

'greenfield' conditions are assumed, i.e. no building is modelled. To compare the experimental and numerical results, the maximum vertical settlement for each value of V_L is assumed as reference. The nonlinear elastic model provides a better fit of the experimental curves, producing a settlement trough narrower than the one predicted by the linear elastic model. Since damage to buildings is proportional to the angular distortion imposed by the settlement trough (Skempton and MacDonald, 1956), accurate modelling of the trough shape and width is important for the evaluation of soil-interaction effects. The nonlinear elastic soil model has been therefore selected to perform the subsequent analyses.

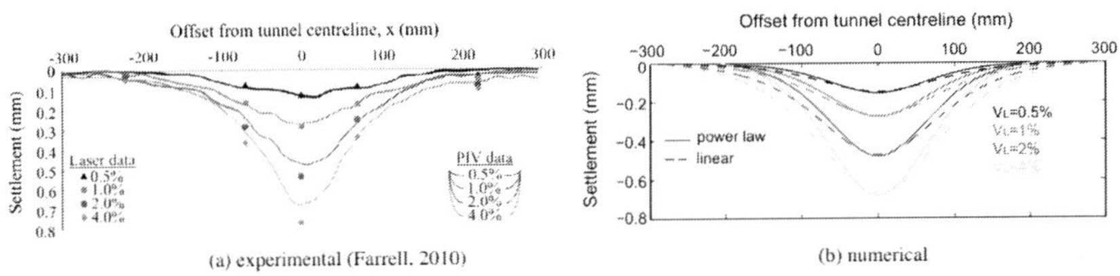

(a) experimental (Farrell, 2010) (b) numerical

Fig. 4. Greenfield vertical displacements of the soil surface.

Particle Image Velocimetry (PIV) measurements of the greenfield vertical and horizontal displacements of the soil (Fig. 5a) were also used to evaluate the performance of the selected soil model. Although the simplified assumption of neglecting a failure criterion for the soil does not allow the model to capture the failure mechanism propagating up from the tunnel crown at high volume losses (Fig. 5a), the model is able to simulate the gradual decrease of settlement towards the ground surface. Also the distribution of horizontal settlements within the soil is adequately simulated by the computational model (Fig. 6). The model does reproduce the experimental concentration of the displacements between the tunnel shoulders and the soil surface for increasing volume loss, with larger displacements at the tunnel shoulders and soil surface and negligible displacements above the tunnel centreline. Note that at volume loss $V_L = 0.5\%$ the displacements measured experimentally were close to the limit precision of PIV (0.01 mm (Farrell, 2010)), making the top pair of figures in Fig. 6 less suitable for direct comparison.

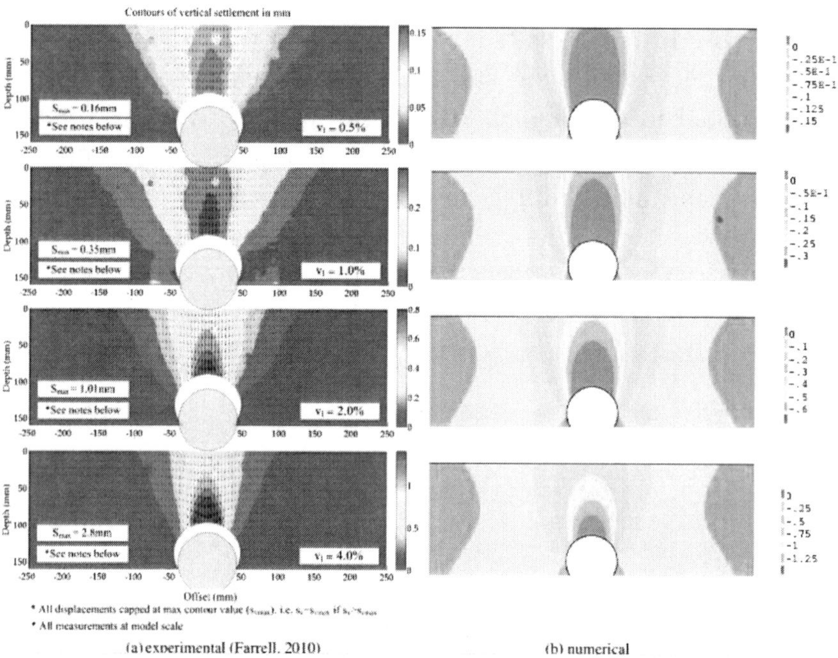

Fig. 5. Contour plot of soil vertical displacements, greenfield case.

3.2. Soil–structure interaction tests

Fig. 7 compares the experimental and numerical displacements of the most flexible (STR-1) and stiffest (STR-4) beams in the vertical direction. The numerical curves, labelled with increasing volume losses to facilitate the visual comparison, correspond to the same values of tunnel radial pressure which produced the greenfield curves in Fig. 4b (also shown in Fig. 7a). The numerical model correctly reproduces both the behaviour of the flexible beam, which follows closely the greenfield deformations, and the stiff beam, which exhibits a significantly smaller deflection. The vertical

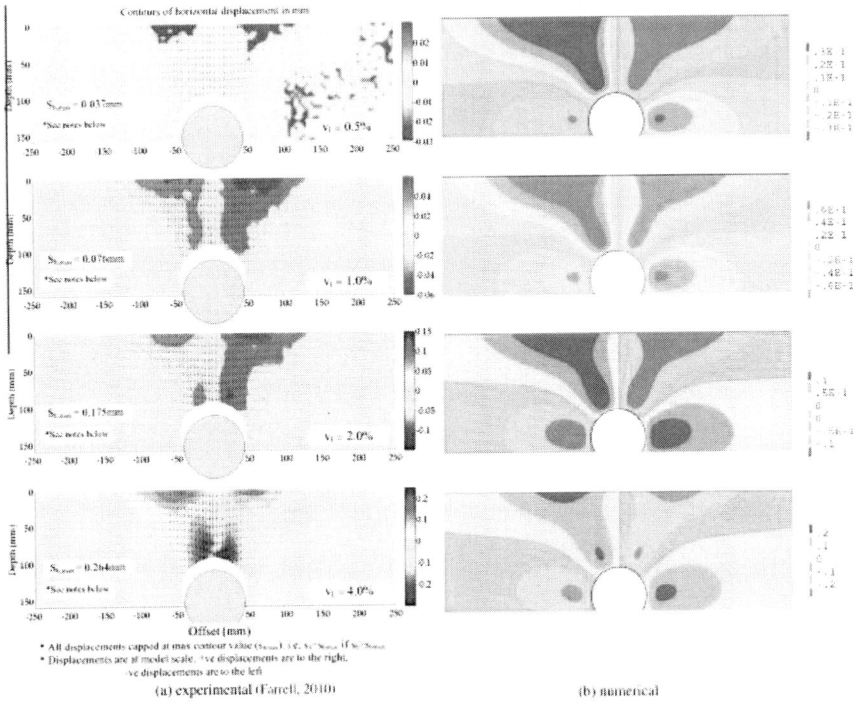

Fig. 6. Contour plot of soil horizontal displacements, greenfield case.

displacement of the soil surface is shown in Fig. 8. Figs. 2a and 8a show a significant difference between the vertical displacements of STR-4 and the settlements of the soil beneath the structure; this difference could not be explained without taking into account the formation of a gap between the soil and the structure base. The key element for an accurate prediction of the experimental results is therefore the use of a no-tension interface between the soil and the beam. Fig. 7b shows that without the interface the structure would be pulled by the subsiding soil, reaching a final deformation of intermediate magnitude between the experimental test and the greenfield case. This interpretation is supported by the experimental measurements of the soil–structure gap (Fig. 9a). The model with the interface captures the increase of the gap as the beam stiffness and the volume loss increase, and provides a good prediction of the gap values for all the analysed structures (Fig. 9b).

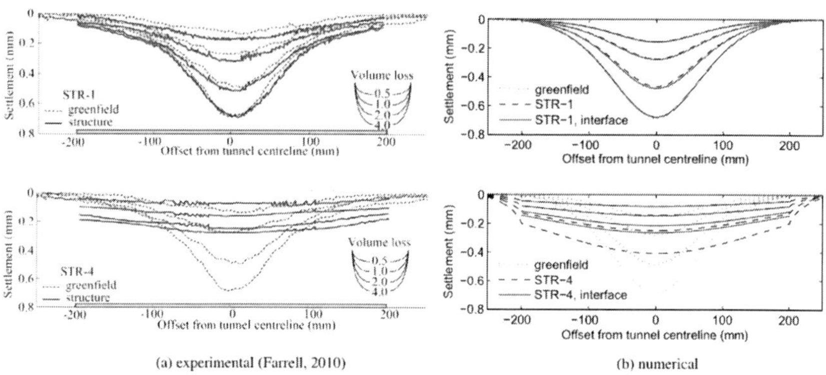

Fig. 7. Vertical displacements of aluminium beam structures.

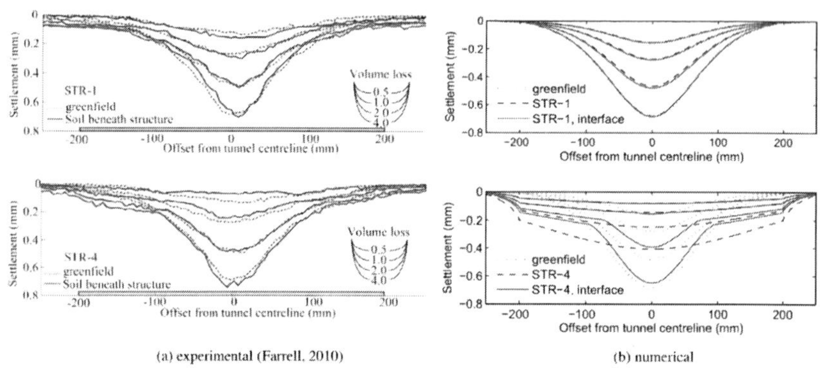

Fig. 8. Vertical displacements of soil.

Horizontal displacements of the beam and soil surface are presented in Fig. 10 and Fig. 11. Consistent with field data (Mair, 2003), the structures do not develop significant horizontal displacements (Fig. 10), and therefore negligible horizontal strains arise. Furthermore, the structures act as a partial restraint to the soil, which exhibits horizontal displacements either smaller than or equal to the greenfield case (Fig. 11). As for the vertical displacements of the beam, the response of the soil in the horizontal direction is

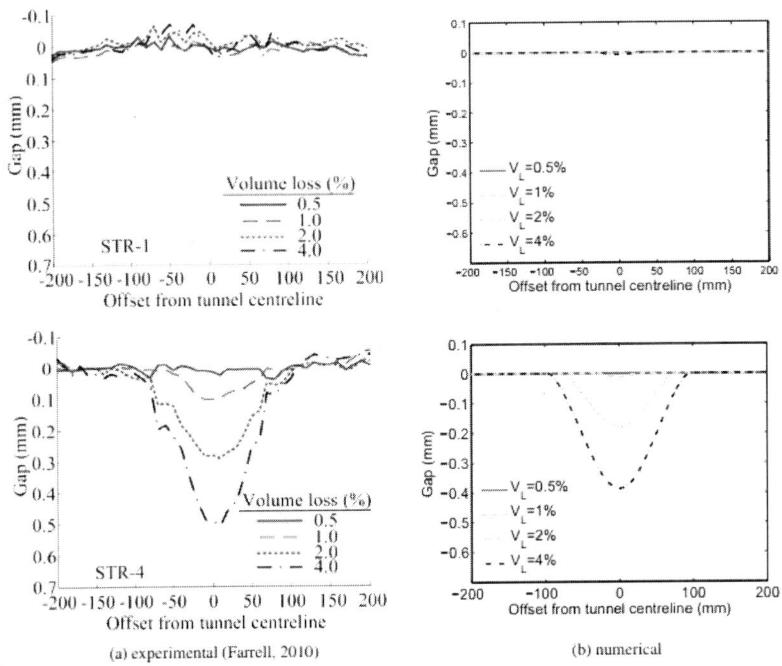

Fig. 9. Soil–structure gap.

connected with the soil–structure gap. For STR-4 and higher values of volume loss, the soil–structure gap allows the maximum soil displacements to nearly match the greenfield results. However, at greater distances from the tunnel centreline, the gap closes and the horizontal soil displacements are negligible (Fig. 11a). For all volume losses and structures the numerical model produces profiles in reasonable agreement with the experimental curves, with a general tendency to overestimate the displacements close to the area of the beam where the gap is formed.

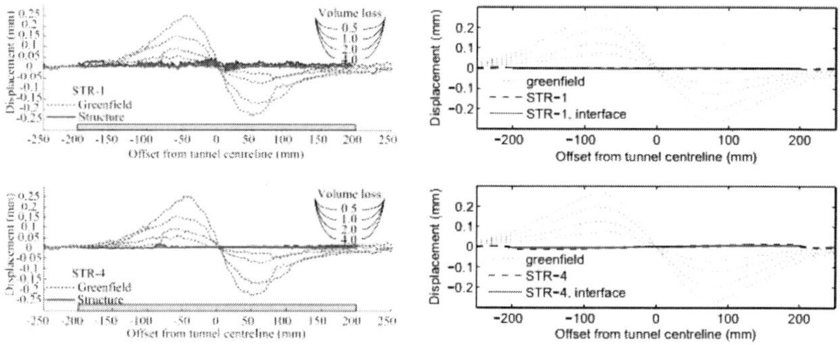

(a) experimental (Farrell, 2010) (b) numerical

Fig. 10. Horizontal displacements of aluminium beam structures.

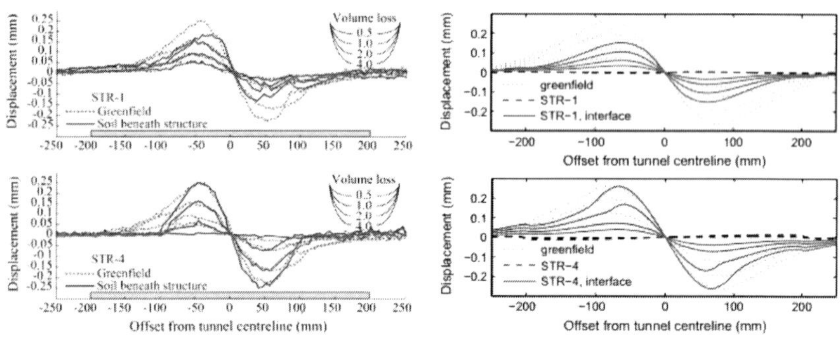

(a) experimental (Farrell, 2010) (b) numerical

Fig. 11. Horizontal displacements of soil.

The analyses of the soil vertical displacements with and without surface structures (Fig. 12) confirm the capability of the model to reproduce the increase in settlement trough width due to the presence of the aluminium beam. The effect is particularly evident when comparing the greenfield case with the STR-4 results (first and fourth rows of Fig. 12, respectively).

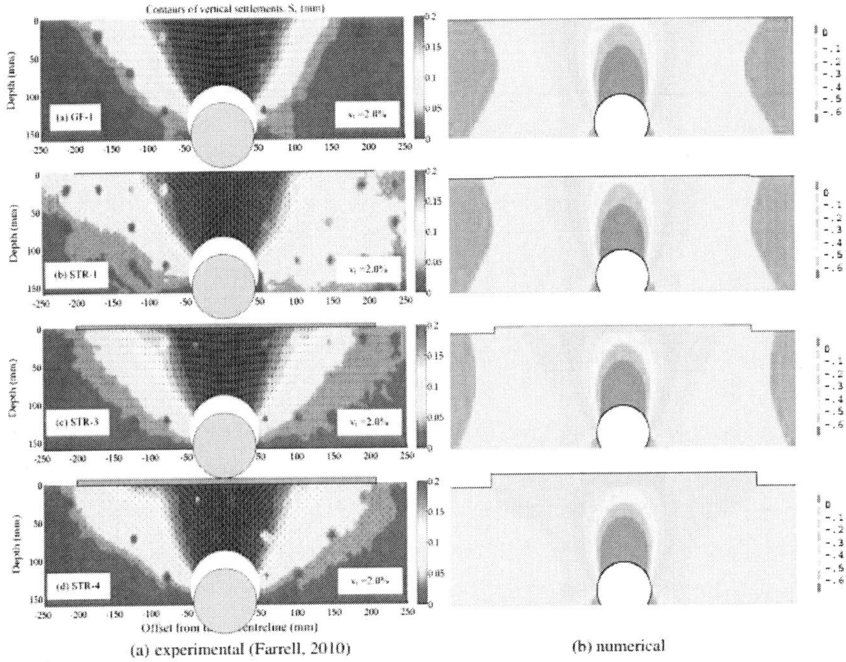

Fig. 12. Contour plot of soil vertical displacements. From top to bottom: greenfield case, STR-1, STR-3 and STR-4.

As mentioned in Section 3.1, the nonlinear elastic model captures the surface displacements with acceptable accuracy up to realistic values of volume loss, after which the failure mechanism propagating from the tunnel crown would govern the settlement profile. Similarly, the model captures the interaction between the soil and surface structures on shallow foundations, but effective modelling of buildings on pile foundations would likely require inclusion of a soil failure mechanism. Additionally, the linear elastic assumption is appropriate for aluminium beams and for identifying when cracking will occur, but modelling the post-cracking response of an existing structure would require a nonlinear material model.

3.3. Modification factors

In an extension of the assessment method by Burland and Wroth (1974), based on the concept of critical strain, Potts and Addenbrooke (1997) measured the effect of the soil–structure interaction on the beam deflection by using the modification factors:

$$M_{DR,sag} = \frac{(\Delta_{sag}/L_{sag})}{(\Delta_{sag}/L_{sag})_{gr}} \qquad M_{DR,hog} = \frac{(\Delta_{hog}/L_{hog})}{(\Delta_{hog}/L_{hog})_{gr}} \tag{3}$$

where L_{sag} and L_{hog} are the building length in the concave (sagging) and convex (hogging) part of the greenfield settlement profile, respectively, Δ_{sag}/L_{sag} andΔ_{hog}/L_{hog} are the corresponding ratios between the relative deflection Δ , defined as the maximum displacement relative to the straight line connecting the two ends of the sagging and hogging part of the building base, and the respective building length, while $(\Delta_{sag}/L_{sag})gr$ and $(\Delta_{hog}/L_{hog})gr$ are the greenfield deflection ratios that would occur without the structure. Potts and Addenbrooke proposed design charts to associate the modification factor to specific features of the building and the soil, summarised in the relative bending stiffness:

$$\rho^* = \frac{EI}{E_s(L/2)^4} \tag{4}$$

where E is the Young's modulus, I is the second moment of area of the cross section,L is the total building length and E_s is the soil characteristic stiffness. Franzius et al. (2006) modified the relative stiffness, which was originally defined in plane strain conditions, to make it dimensionless in both two and three dimensions, by including the effect of the tunnel depth z_0 and the building width B:

$$\rho^*_{mod} = \frac{EI}{E_s L^2 z_0 B} \tag{5}$$

Based on experimental tests and field data, Farrell (2010) partitioned the relative bending stiffness in the sagging and hogging zone of the greenfield settlement profile curvature:

$$\rho^*_{\text{sag.par}} = \frac{EI}{E_s L^3_{\text{sag}} B} \quad \rho^*_{\text{hog.par}} = \frac{EI}{E_s L^3_{\text{hog}} B} \tag{6}$$

The modification factor for the deflection ratio has been used to complete the validation of the numerical model in terms of global structural deformation. Fig. 13 shows the modification factors versus the relative bending stiffness as defined in Eq. (5), while Fig. 14 shows the same but for the relative bending stiffness values defined by Eq. (6). Note that in the experimental graph the soil stiffness is assumed to be dependent on the volume loss and therefore the relative stiffness increases for increasing values of volume loss (Farrell, 2010). In the numerical model, a single reference value for the soil stiffness is assumed throughout the whole analysis. For each structure, the numerical results show a range of modification factors comparable to the experimental values.

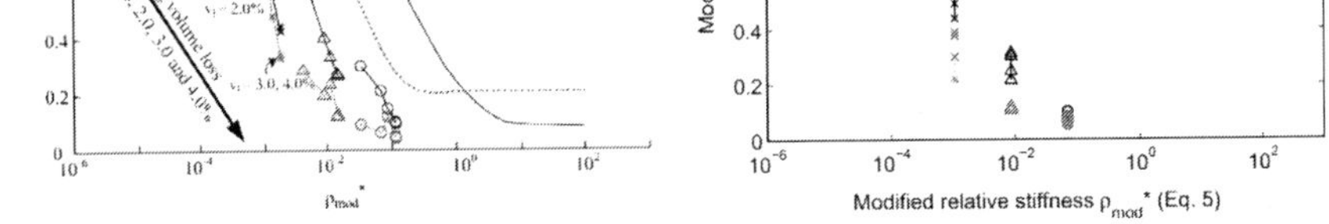

(a) experimental (Farrell, 2010) (b) numerical

Fig. 13. Deflection ratio modification factor vs modified relative bending stiffness (Eq. (5)).

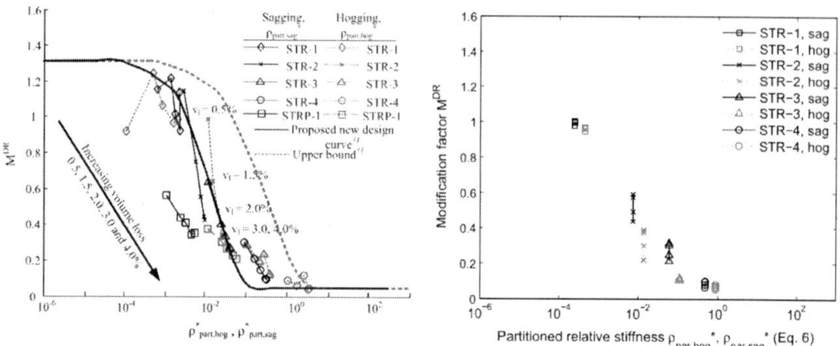

(a) experimental (Farrell, 2010)

(b) numerical

Fig. 14. Deflection ratio modification factor vs partitioned modified relative bending stiffness (Eq. (6)).

For small values of relative stiffness, the structure deformations are generally similar to the greenfield ones and therefore the modification factor MDR is close to unity, while for increasing values of relative stiffness the beam deflection ratio reduces and so doesMDR. With respect to the experimental results, the numerical model tends to underestimate the modification factors for the most flexible structure and for the lowest value of volume loss. These are the cases where the beam distortion approximates to the greenfield one, sometimes showing displacements even larger than in the greenfield case (Fig. 7a, STR-1). The main reason for this is the embedding of the ends of the structure into the soil, a mechanism which is not captured by the numerical model. However, the results provide a generally good estimation of the modification factors over a wide range of structure stiffness. This offers a reliable base to perform a sensitivity study on how the building weight influences the soil–structure interaction mechanism.

4. EFFECT OF BUILDING WEIGHT

The results above indicates the presence of a gap between the soil and the structure in the case of stiffer structures and higher volume loss values. This section investigates how that gap, and the soil–

structure interaction more generally, may be affected by the weight of the structure, which is typically ignored in assessment procedures. The building stiffness and weight variations selected are first presented, followed by the results of the parametric study. Differences with results from previous studies are highlighted and discussed.

4.1. Parametric variations

Fig. 15a gives a graphical overview of the numerically tested structures in terms of bending and axial stiffness at prototype scale. The variations include structures STR-1 to STR-4 from Section 3, as well as new structures STR-5 to STR-8. In order to relate these variations to real buildings, data from buildings monitored during the Jubilee Line Extension (Mair and Taylor, 2001 and Dimmock and Mair, 2008) are also shown in the graph. The selected stiffness values represent an upper and lower envelope for the real stiffnesses, particularly for the bending stiffness when calculated with the methodologies described in Mair and Taylor, 2001 and Dimmock and Mair, 2008. The exact stiffness values assumed in the parametric study are reported in Table 2.

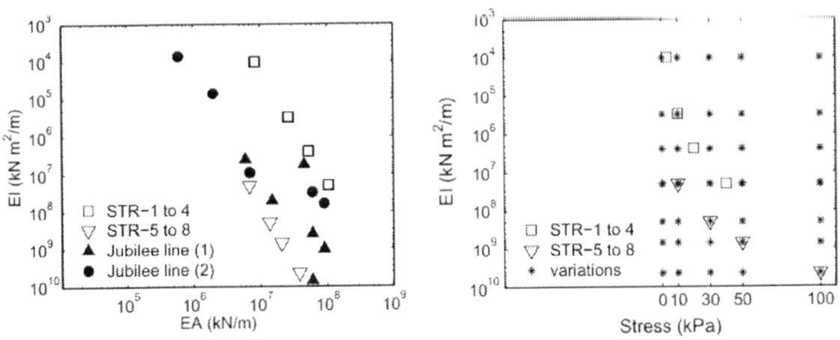

(a) Bending (EI) vs axial (EA) stiffness for numerical variations and field data

(1) EI and EA according to Mair and Taylor (2001)

(2) EI and EA according to Dimmock and Mair (2008)

(b) Base stress vs bending stiffness for numerical variations

Fig. 15. Bending stiffness, axial stiffness and base stress at the prototype scale for the reference centrifuge models, real buildings and the parametric variations performed in this study.

Table 2. Parametric variations of bending and axial stiffness combination.

	Axial stiffness *EA* (kN/m)	Bending stiffness *EI* (kN m)		Axial stiffness *EA* (kN/m)	Bending stiffness *EI* (kN m)
STR-1	8.4×10^6	1.0×10^4	STR-5	6.9×10^6	2.0×10^7
STR-2	2.6×10^7	3.0×10^5	STR-6	1.4×10^7	2.0×10^8
STR-3	5.2×10^7	2.4×10^6	STR-7	2.1×10^7	6.0×10^8
STR-4	1.0×10^8	2.0×10^7	STR-8	3.8×10^7	4.4×10^9

Four different values of weight were applied to each of the analysed structures. These values approximate to the stresses acting at the base of real buildings of 1, 3, 5 and 10 storeys (Franzius et al., 2004) and they are listed in Table 3. Fig. 15b shows the complete set of variations in terms of bending stiffness and base stresses.

Table 3. Parametric variations of weight.

	Base stress (kN/m^2)
σ_1	10
σ_2	30
σ_3	50
σ_4	100

4.2. Results

Fig. 16 shows the effect of the weight on the deflection of STR-1 and STR-4 for selected volume losses. The results indicate that the vertical displacements of the beam increase with increasing applied weight. For higher volume losses this effect can be substantial: in the range of assumed weights, a significant loading can increase the beam settlements up to 70%. For relatively stiff structures, the weight influences the embedding of the structure into the soil, which is indicated by the vertical displacements at the two ends of the beam (Fig. 16d) and the development of the gap between the soil and the beam base in the case of the stiffer structure STR-4 (Fig. 17); Fig.

17 shows how the increase in weight tends to make the structure STR-4 follow the soil displacements and therefore reduces the gap.

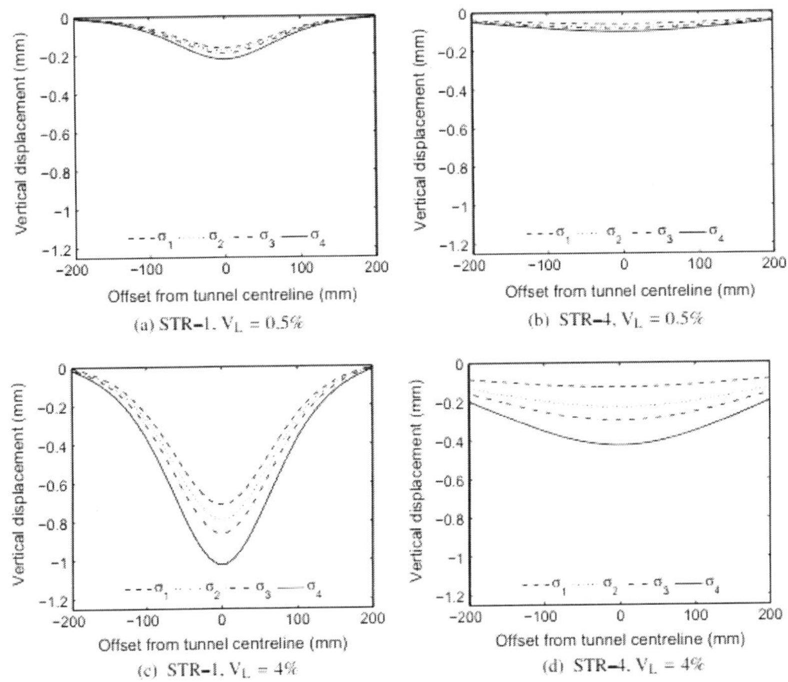

Fig. 16. Weight influence on the vertical displacements of selected structures (STR-1 and STR-4).

The modification factors derived from the numerical analysis for all the structures, for a representative volume loss of 0.5% are plotted in Fig. 18a and b as a function of the partitioned relative stiffness in sagging and hogging, respectively (Eq. (6)). The results indicate that higher loading can lead to an increased modification factor, suggesting that the weight of the structure can reduce the effect of an increment in relative stiffness.

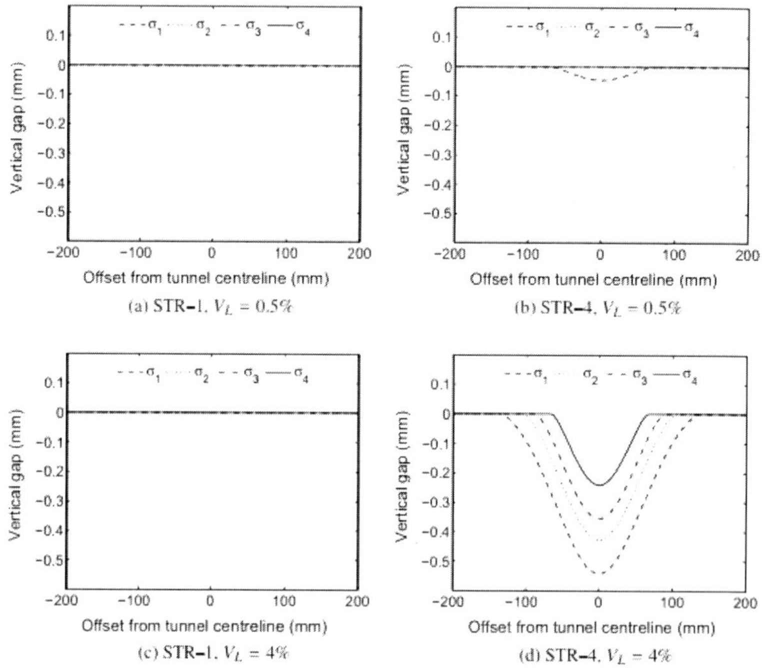

Fig. 17. Weight influence on the gap between the soil and selected structures (STR-1 and STR-4).

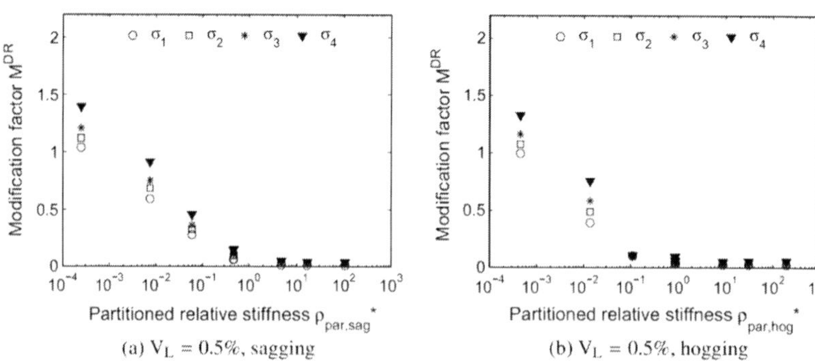

Fig. 18. Deflection ratio modification factor vs relative bending stiffness (Eq. (6)), all variations.

The impact of the weight on the building response is particularly significant when considering that this factor is neglected in the empirical–analytical procedures currently used for the assessment of potential damage. As introduced in Section 1, these procedures are based on the analyses performed on weightless beam models of buildings (Burland and Wroth, 1974 and Potts and Addenbrooke, 1997). Later numerical analyses (Franzius et al., 2004) assessed the weight as a negligible factor if compared with the corresponding realistic variations of bending stiffness values. These analyses (Franzius et al., 2004), although including a comprehensive evaluation of variation of weight, consider a smaller range of relative stiffness values and were derived from a numerical model which neglected the possibility of a gap between the soil and the structure. The assumption of no gap was numerically evaluated by verifying the absence of normal tensile forces at the soil–beam interface for a singular case of stiffness and weight. While this check seems reasonable, the experimental data (Farrell, 2010) show that the structural response can involve the formation of a soil–structure gap. It should be noted that the values of relative stiffness and weight simulated by Farrell (2010) are within the realistic range of values for existing buildings. In addition, the numerical analyses described in this paper, characterised by the specific simulation of the soil-interface gap, suggest the relevance of the weight in the experimentally observed mechanism by including a wider range of stiffness values. The need to include the building weight in 2D and 3D numerical modelling of tunnelling has also been stated byLiu et al., 2000 and Rampello et al., 2012, respectively.

Fig. 18 highlights that the impact of the building weight on the modification factor decreases with the increase in relative stiffness. This suggests a possibility to correlate the modification factors with an indicator that combines the relative stiffness and weight of a structure. For example, a normalised relative stiffness ρ_{norm} can be defined in sagging and hogging as:

$$\rho_{norm.sag} = \frac{\rho^*_{sag.par}}{\frac{\sigma}{\sigma_1}} \qquad \rho_{norm.hog} = \frac{\rho^*_{hog.par}}{\frac{\sigma}{\sigma_1}} \tag{7}$$

where σ is the total bearing pressure acting at the building foundations and σ_1 is the bearing pressure produced by one typical storey of the same building.

For cases where the building weight and bearing pressure are directly correlated with the number of storeys, like in this study (Table 2), the dimensionless coefficient $\frac{\sigma}{\sigma_1}$ corresponds to the number of storeys n_s:

$$\rho_{norm.sag} = \frac{\rho^*_{sag.par}}{n_s} \qquad \rho_{norm.hog} = \frac{\rho^*_{hog.par}}{n_s} \tag{8}$$

In Fig. 19, the modification factors are shown as a function of ρ_{norm}, again for a volume loss of 0.5%. The modification factors tend to align, identifying a decrease with the increase of the normalised relative stiffness. Similar results were found for higher values of V_L. In Fig. 20, the numerical results in terms of modification factors are plotted first against the partitioned relative stiffness (Eq. (6)) and then against the normalised relative stiffness (Eq. (8)) for all values of volume loss.

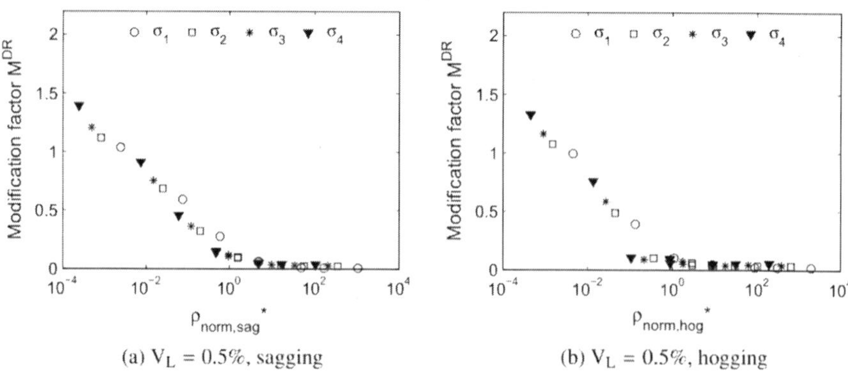

Fig. 19. Deflection ratio modification factor vs normalised relative bending stiffness, all variations.

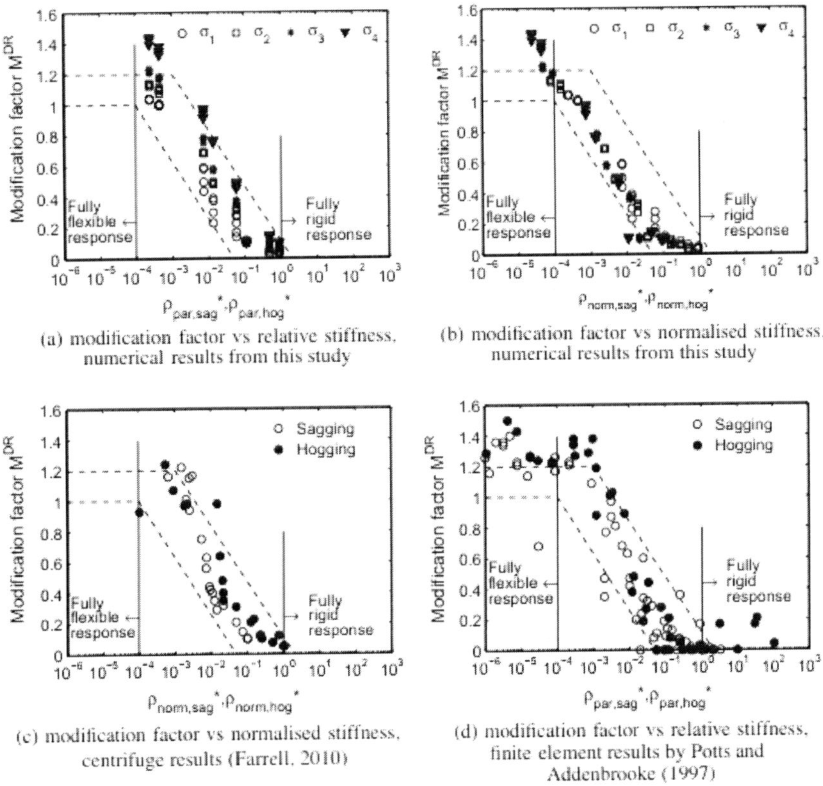

(a) modification factor vs relative stiffness, numerical results from this study

(b) modification factor vs normalised stiffness, numerical results from this study

(c) modification factor vs normalised stiffness, centrifuge results (Farrell, 2010)

(d) modification factor vs relative stiffness, finite element results by Potts and Addenbrooke (1997)

Fig. 20. Deflection ratio modification factor vs relative and normalised bending stiffness: comparison with centrifuge, numerical and field data. The dotted lines define the envelope identified by Mair (2013).

When the weight is included in the building characterisation, the results fall into a narrower band within the envelope identified by Mair (2013) which applies to centrifuge models and field case histories. For values of $\rho_{par}^{*} < 10^{-4}$ and $\rho_{par}^{*} > 1$, the structure behaves as fully flexible and fully rigidly, respectively. For this reason, only the numerical results corresponding to $10^{-4} < \rho_{par}^{*} < 1$ are included in Fig. 20a and b. The analysis of the centrifuge data from Farrell (2010) (Fig. 20c) and the numerical results from Potts and Addenbrooke (1997) (Fig. 20d) is consistent with this. It should be noted that the weight effect is included only for structures with weight corresponding to one or more storeys. The coefficient n_s in Eq. (8) is assumed equal to 1 in the other cases. For the weightless beams analysed by Potts and Addenbrooke, the normalised stiffness therefore coincides with the relative stiffness.

5. CONCLUSIONS

In this paper, the interaction between tunnelling in sand and surface structures has been numerically investigated. A 2D finite element model with nonlinear interface elements was used to investigate the role of the soil–structure gap in modifying the soil greenfield settlement and building response. The comparison with the results of centrifuge tests showed that the model effectively reproduces the experimentally measured soil and structure displacements and that the use of a no-tension interface is necessary to simulate the soil–structure gap and the building deformation in the case of stiffer buildings.

The numerical model was then used to perform a sensitivity study on the interaction between building stiffness and weight. The results show that the building response depends not only on the relative stiffness between the structure and the soil but also on the building weight, which is typically neglected in the current assessment procedures. The presentation of results in terms of relative stiffness normalised to a dimensionless indicator of the building weight demonstrated the potential value of including the weight in the damage assessment and was shown to be consistent with available experimental and numerical observations.

ACKNOWLEDGEMENTS

Financial support was provided by the Engineering and Physical Sciences Research Council of the United Kingdom, under Grant reference number EP/K018221/1. The research materials supporting this publication can be accessed at https://www.repository.cam.ac.uk/handle/1810/249085. The authors would like to acknowledge Dr. Ruaidhri Farrell for providing experimental data.

REFERENCES

1. Amorosi, D. Boldini, G. De Felice, M. Malena, M. Sebastianelli **Tunnelling-induced deformation and damage on historical masonry structures** Geotechnique, 64 (2) (2014), pp. 118–130
2. M. Boonpichetvong, J.G. Rots **Settlement damage of masonry buildings in soft-ground tunneling** Struct. Eng., 83 (1) (2005), pp. 32–37
3. M.D. Boscardin, E.J. Cording **Building response to excavation-induced settlement** J. Geotech. Eng., 115 (1) (1989), pp. 1–21
4. BRE, 1995. Digest 251: Assessment of Damage in Low Rise Buildings with Particular Reference to Progressive Foundation Movements. Building Research Establishment, Tech. rep. Building Research Establishment.
5. H.J. Burd, G.T. Houlsby, C.E. Augarde, L. G **Modelling tunnelling-induced settlement of masonry buildings** Proc. Inst. Civil. Eng.: Geotech. Eng., 143 (1) (2000), pp. 17–29
6. J.B. Burland **Assessment of risk of damage to buildings due to tunneling and excavation** K. Ishihara (Ed.), Proceedings of the 1st International Conference on Earthquake Geotechnical Engineering, A.A. Balkema, IS, Tokyo, Japan (1995), pp. 1189–1201
7. J.B. Burland, C.P. Wroth **Settlement of buildings and associated damage** Proceedings of Conference on Settlement of Structures, Pentech Press, Cambridge (1974), pp. 611–654

8. Burland, J.B., Broms, B.B., de Mello, V.F.B., 1977. Behaviour of foundations and structures. In: Proceedings of the 9th International Conference on Soil Mechanics and Foundation Engineering, vol. 2, 1977, pp. 495–546.

9. J. Clarke, D.F. Laefer **Evaluation of risk assessment procedures for buildings adjacent to tunnelling works** Tunnel. Under. Space Technol., 40 (2014), pp. 333–342

10. M.J. DeJong, M.A.N. Hendriks, J.G. Rots **Sequentially linear analysis of fracture under non-proportional loading** Eng. Fract. Mech., 75 (18) (2008), pp. 5042–5056

11. M. Devriendt, E. Palmer, R. Hill, D. Lazarus **Historic and non-historic building impact assessment methodology for major tunnelling infrastructure projects** C. Viggiani (Ed.), Geotechnical Engineering for the Preservation of Monuments and Historic Sites, CRC Press (2013), pp. 335–341

12. P.S. Dimmock, R.J. Mair **Effect of building stiffness on tunnelling-induced ground movement** Tunnel. Under. Space Technol., 23 (2008), pp. 438–450

13. Farrell, R.P., 2010. Tunnelling in Sands and the Response of Buildings, Ph.D. thesis. University of Cambridge.

14. R.P. Farrell, R.J. Mair **Centrifuge modelling of the response of buildings to tunneling** ,in: L. Seward (Ed.), Physical Modelling in Geotechnics, Proceedings of the 7th International Conference on Physical Modelling in Geotechnics, Vol. 2, , CRC Press, Zurich, Switzerland (2011), pp. 549–554

15. Franzius, J.N., 2003. Behaviour of Buildings Due to Tunnel Induced Subsidence, Ph.D. thesis. Imperial College, London.

16. J.N. Franzius, D.M. Potts, T.I. Addenbrooke, J.B. Burland **The influence of building weight on tunnelling-induced ground and building deformation** Soil Found., 44 (1) (2004), pp. 25–38

17. J.N. Franzius, D.M. Potts, J.B. Burland **The response of surface structures to tunnel construction** Proc. Inst. Civil. Eng.: Geotech. Eng., 159 (1) (2006), pp. 3–17

18. J.N. Franzius, D.M. Potts, J.B. Burland **Twist behaviour of buildings due to tunnel induced ground movement** E. Kwast, K. Bakker, W. Broere, A. Bezuijen (Eds.), Proceedings of the 5th International Conference on Geotechnical Aspects of Underground Construction in Soft Ground, Taylor and Francis, Amsterdam (2006), pp. 107–113

19. G. Giardina, M.A.N. Hendriks, J.G. Rots **Numerical analysis of tunnelling effects on masonry buildings: the influence of tunnel location on damage assessment** Adv. Mater. Res., 133 (2010), pp. 289–294

20. G. Giardina, A. Marini, M.A.N. Hendriks, J.G. Rots, F. Rizzardini, E. Giuriani **Experimental analysis of a masonry façade subject to tunnelling-induced settlement** Eng. Struct., 45 (2012), pp. 421–434

21. G. Giardina, A.v.d. Graaf, M.A.N. Hendriks, J.G. Rots, A. Marini **Numerical analysis of a masonry façade subject to tunnelling-induced settlement** Eng. Struct., 54 (2013), pp. 234–247

22. G. Giardina, M.A.N. Hendriks, J.G. Rots **Sensitivity study on tunnelling induced damage to a masonry façade** Eng. Struct., 89 (2015), pp. 111–129

23. G. Giardina, M.A.N. Hendriks, J.G. Rots **Damage functions for the vulnerability assessment of masonry buildings subjected to tunneling** J. Struct. Eng., 141 (9) (2015)

24. K.H. Goh, R.J. Mair **Building damage assessment for deep excavations in Singapore and the influence of building stiffness** Geotech. Eng. J. SEAGS AGSSEA, 42 (3) (2011)

25. V. Guglielmetti, P. Grasso, A. Mahtab, S. Xu (Eds.), Mechanized Tunnelling in Urban Areas: Design Methodology and Construction Control, Taylor and Francis (2008)

26. D.F. Laefer, L.T. Hong, A. Erkal, J.H. Long, E.J. Cording **Manufacturing, assembly, and testing of scaled, historic masonry for one-gravity, pseudo-static, soil–structure experiments** Construct. Build. Mater., 25 (12) (2011), pp. 4362–4373

27. G. Liu, G.T. Houlsby, C.E. Augarde **2-dimensional analysis of settlement damage to masonry buildings caused by tunneling** Struct. Eng., 79 (1) (2000), pp. 19–25

28. N. Losacco, A. Burghignoli, L. Callisto **Uncoupled evaluation of the structural damage induced by tunneling** Géotechnique, 64 (2014), pp. 646–656 10

29. Mair, R.J., 2003. Research on tunnelling-induced ground movements and their effects on buildings – lessons from the Jubilee Line Extension, keynote lecture. In: Jardine, F.M. (Ed.), Proceedings of International Conference on Response of Buildings to Excavation-induced Ground Movements, CIRIA Special Publication 199, RP620. Imperial College, London, UK, pp. 3–26.

30. R.J. Mair **Tunnelling and deep excavations: ground movements and their effects** A.E.A. Anagnostopoulos (Ed.), Proceedings of the 15th European Conference on Soil Mechanics and Geotechnical Engineering – Geotechnics of Hard Soils – Weak Rocks (Part 4), IOS Press (2013), pp. 39–70

31. Mair, R.J., Taylor, R.N., 2001. Building response to tunnelling – case studies from construction of the Jubilee Line Extension, vol. 1. CIRIA special publication 200, London, 2001 (Ch. Elizabeth House: settlement predictions, pp. 195–215).

32. R.J. Mair, R.N. Taylor, J.B. Burland **Prediction of ground movements and assessment of risk of building damage due to bored tunneling** R.J. Mair, R.N. Taylor (Eds.), Proceedings of the International Symposium on Geotechnical Aspects of Underground Construction in Soft Ground, Balkema, Rotterdam (1996), pp. 713–718

33. Melis, M., Rodriguez Ortiz, J., 2001. Consideration of the stiffness of buildings in the estimation of subsidence damage by EPB tunnelling in the Madrid subway. In: Response of Buildings to Excavation Induced Ground Movements Conference. London.

34. E. Namazi, H. Mohamad **Assessment of building damage induced by three-dimensional ground movements** J. Geotech. Geoenviron. Eng., 139 (4) (2013), pp. 608–618

35. Netzel, H.D., 2009. Building Response Due to Ground Movements, Ph.D. thesis. Delft University of Technology.

36. H.L. Nghiem, M. Al Heib, F. Emeriault **Physical model for damage prediction in structures due to underground excavations** Tunnel. Under. Construct. (2014), pp. 155–164

37. J.A. Pickhaver, H.J. Burd, G.T. Houlsby **An equivalent beam method to model masonry buildings in 3D finite element analysis** Comp. Struct., 88 (19–20) (2010), pp. 1049–1063

38. D.M. Potts, T.I. Addenbrooke **A structure's influence on tunnelling-induced ground movements** Proc. Inst. Civil Eng.: Geotech. Eng., 125 (2) (1997), pp. 109–125

39. S. Rampello, L. Callisto, G. Viggiani, F.M. Soccodato **Evaluating the effects of tunnelling on historical buildings: the example of a new subway in Rome/ Auswertung der Auswirkungen des Tunnelbaus auf historische Gebäude am Beispiel einer neuen U-Bahnlinie in Rom** Geomech. Tunnel., 5 (3) (2012), pp. 275–299

40. Rots, J.G., 2000. Settlement damage predictions for masonry. In: Verhoef, L., Wittman, F. (Eds.), Maintenance and Restrengthening of Materials and Structures – Brick and Brickwork. Proceedings of the International Workshop on Urban Heritage and Building Maintenance. Aedificatio, Freiburg, pp. 47–62.

41. A.W. Skempton, D.H. MacDonald **The allowable settlements of buildings** Inst. Civil Eng., 5 (1956), pp. 727–784

42. M. Son, E.J. Cording **Estimation of building damage due to excavation-induced ground movements** J. Geotech. Geoenviron. Eng., 131 (2) (2005), pp. 162–177

43. M. Son, E.J. Cording **Evaluation of building stiffness for building response analysis to excavation-induced ground movements** J. Geotech. Geoenviron. Eng., 133 (8) (2007), pp. 995–1002

7
CHAPTER

Experimental Study of Internal Flow Noise Measurement by Use of a Suction Type Low Noise Wind Tunnel

Yoshifumi Yokoi[1]

[1] National Defense Academy of Japan, Japan

1. INTRODUCTION

The measurement experiment of the fluid-dynamic noise made from the object placed into the air flow is performed using a low noise wind tunnel, a silent airflow wind tunnel, etc. In the low noise wind tunnel, the measures against silence are taken so that the noise generated with a fan or a compressor may not propagate as much as possible to a wind tunnel test section by an air current. As for the surroundings of the test section of a low noise wind tunnel, acoustic free space is provided. Generally a wind tunnel is classified by the form of the channel of a wind tunnel (blow type, suction type and circulating type), the form of the measurement section (open, half-open and sealed), and the existence of circulation of flow. And the practical wind tunnels are classified into 13 kinds (Mochizuki & Maruta, 1996). Figure 1 illustrates the circulation environment for the airflow between the blower and the measurement section, the types of duct in the wind tunnel (blow, suction and circulating) and the types of measurement section (open, half-open and sealed). The merit of each type of the wind tunnel and the weak point are summarized as follows. In the merit of the blow type, the composition is simple and small the installation space. In the liberating measurement section of jet-type, the usage of use becomes various. The week point is to need big power because the pressure loss is large. Flowing quantity will come to receive the fluctuation easily in turbulence. The measurement section is that the temperature raises more than the temperatures of air in the surrounding. The merit of the suction-type should be able to be composed the rectification part short, and more compactly than the blow-type. The temperature of the measurement section is the same as the temperature of the space in the surrounding. The weak point is to receive the influence of the fluctuation of the outer air flow large. The measurement section must become negative pressure from the atmospheric pressure. An enough space for the rectification is needed on the suction side. The merit of the circulating blow-type is not to receive turbulence. The experiment on all-round is possible in the open-type measurement section. The weak point is to take time until stabilizing and it be easy to rise in the temperature. Merits of the circulating suction-type are that turbulence is not received and the rectification part is short. The weak point is to take time until stabilizing. A very wide space is necessary forward of the suction mouth. Merits of the circulating type to unnecessary big power and not to receive turbulence. The stability of the flow is also early. Especially, efficiency is very good and the pressure loss is a little in the sealed-type measurement section. The weak point is to need noting in the rise's of the air flow temperature becoming remarkable. The object flow must be limited. A wide installation space is needed. In addition, there are a peculiar merit and a weak point respectively by the measurement section shape, and they are summarized as follows. The merit of the open-type measurement section is that the limitation concerning the size and the shape of the test piece is a little. The weak point is to receive turbulence by the suck of air. Merits of the half-open type measurement section are permitted the test piece diversity and are hard of turbulence to receive. The weak point is that the measurement room becomes negative pressure easily. The merit of the sealed-type measurement section is to become the most efficient wind tunnel, and to hardly receive turbulence. The weak point is to receive the limitation to the size and the shape of the test piece. Among these, it is required that the wind tunnel aiming at measurement of a fluid-dynamic noise secures the acoustic free space of silence and a test section. Moreover, it is also required that the spatial relationship of a test model and a microphone can be set up freely. Therefore, many blow-type wind tunnels with the measurement room and half-open type test section by which sound insulation processing was carried out with the sound-absorbing material are used. On the other hand, use of a microphone is difficult in an air flow, and the measurement technique of a fluid-dynamic noise has not been established. Therefore, the wind tunnel with a sealed type test section can scarcely be seen. Accordingly, measurement of the fluid-dynamic noise of internal flows, such as a flow inside a gas turbine or a jet engine, and a pipeline, a flow of the around of the support in a duct, is not in the state which can be performed immediately. As for the present condition, there are also few examples of verification of measurement of the fluid-dynamic noise of an internal flow. So, it is very important to establish the measurement technique of the fluid-dynamic noise of an internal flow in engineering. In measurement of the fluid-dynamic noise using a low noise wind tunnel, when an open-type test section is used, it is reported that there is a case where it becomes impossible for a back ground noise not to be amplified by the large turbulence produced with the edge of the jet stream from a nozzle, or for generating of the sound which is not a measuring object to be observed by interference of a jet and a model sample, or to maintain the two dimensional characteristic of a flow etc. Moreover, when a sealed type test section is used, on the usual surface of a wall, sound reflects, and exact measurement cannot be performed, but if the material which can bear wind pressure that sound tends to penetrate the surface of a wall is used, it is reported that the sealed type test section will probably be better (Fujita, 1994, 1996).

The purpose of this study is examination of the measurement technique of the fluid-dynamic noise of an internal flow. In this study, it proposes carrying out burial setting of the microphone to the test section equipped with a fibered glass. The suction type low noise wind tunnel with such a test section for verification was created, and measurement of the fluid-dynamic noise made from the circular cylinder placed into the air flow was tried. Comparison examination of the measurement result obtained by this measurement technique was carried out with the measurement result obtained in the blow type wind tunnel. As a result, it was shown that the same characteristic is obtained about the change in a sound pressure level or peak frequency. Moreover, since the target acoustic frequency was caught clearly, it was shown that it is convenient for examination of an acoustic effect. This measurement technique showed clearly that usefulness is high to fluid-dynamic noise measurement of the internal flow.

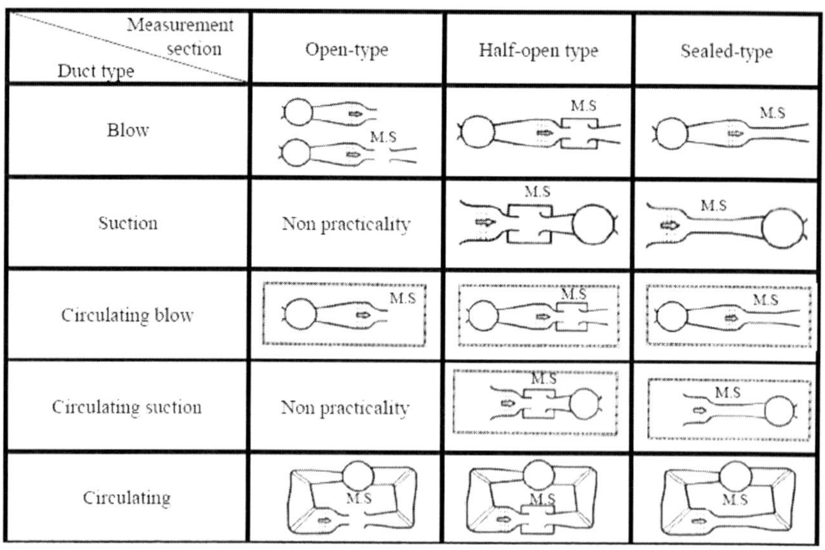

FIGURE 1. Wind tunnel classifications (Mochizuki & Maruta, 1996); the circle represents the blower, the arrow shows direction of the flow, and "M.S" is the measurement section

2. EXPERIMENTAL APPARATUS AND METHOD

This chapter describes the used equipment, a tool, and the procedure of an experiment.

2.1. OUTLINE OF THE EXPERIMENTAL APPARATUS

The experimental apparatus consists of a low noise wind tunnel and measuring equipment. Figure 2shows the schematic diagram of a low noise wind tunnel. The low noise wind tunnel is constituted from the bell mouse, the test section, the silence duct, and the fan by the inhaled type wind tunnel with a sealed type test section. In order to reduce fan generating noise, the inside of a silence duct is divided into four in the shape of a cell, the sound-absorbing material (fibered glass) is stuck on all the surface of a wall, and the fan is installed in the fan room by which interior was carried out with the sound-absorbing material with a silence exhaust port with which three splitter walls were set. Regulation of airflow velocity which passes a test section is performed by carrying out inverter control of the number of rotations of the fan by remote control. A measuring device is divided roughly into fluid-dynamic noise measurement equipment and the air flow velocity measurement equipment. Fluid-dynamic noise measurement equipment consists of directive capacitor microphone (RION, UC-30, hereafter it is called microphone for convenient), precision noise level meter (RION, NA-40), and FFT analyzers (Ono Sokki, CF-5220). The air flow velocity measurement equipment consists of a hot-wire anemometer (DISA, TYPE55) and a digital pressure gauge (Cosmo Instruments, DM-3100B). As for measurement of the turbulence intensity to the flow velocity distribution and a main flow, the hot-wire anemometer was used. The pressure difference between the surface of a wall (static pressure) of a test section and atmospheric pressure was measured with thc digital pressure gauge.

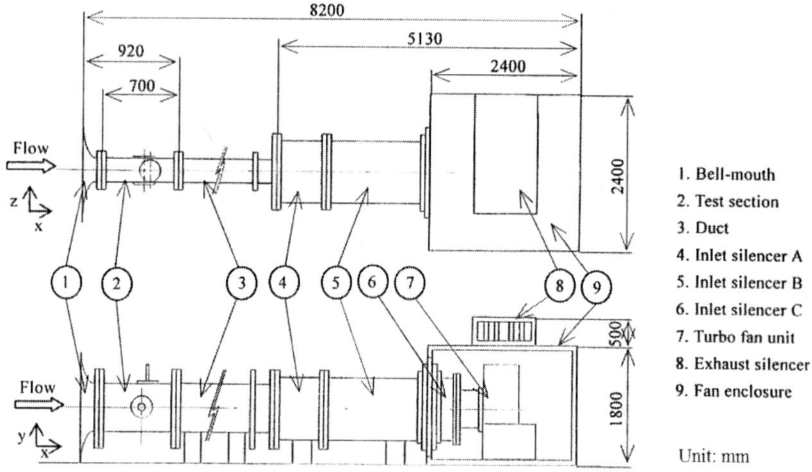

FIGURE 2. The schematic diagram of the wind tunnel

2.2. MEASUREMENT SECTION AND TEST CYLINDERS

Figure 3 shows the schematic diagram of a measurement section (test section). The measurement section is a rectangular cross-section, 376mm (y direction) in height and 160mm (z direction) in width, with both side walls made of a transparent acrylic resin 700mm (x direction) in length, and a board thickness of 10mm. The turntable installation hole with a diameter of 100mm was installed from the edge of the measurement section upstream side to the position at 350mm in the centerline. Upper and lower walls act as the sound absorbing walls (fibered glass walls), with 50mm-thick fibered glass placed on a 15mm-thick transparent acrylic board. Half free space is made in acoustics by installing this sound absorbing wall. The microphone and the hot-wire probe are set up from the edge of the measurement part upstream side to the position at 400mm in the centerline. The surface of microphone and the surface of fibered glass are set at the same level. The hot-wire probe can be moved in a vertical direction in the measurement section (y direction) using the traverse device. The test circular cylinder can be set within a range of 5mm-45mm up from the center of the turntable. Here, the center-to-center spacing of the microphone and the circular cylinder make adjustments within a range of 5mm-95mm possible. The test circular cylinder is made from brass, span length is 160mm and the surface is finished smoothly. The test circular cylinder is with seven kind, and each diameter is 6mm, 10mm, 15mm, 20mm, 25mm, 30mm, and 40mm.

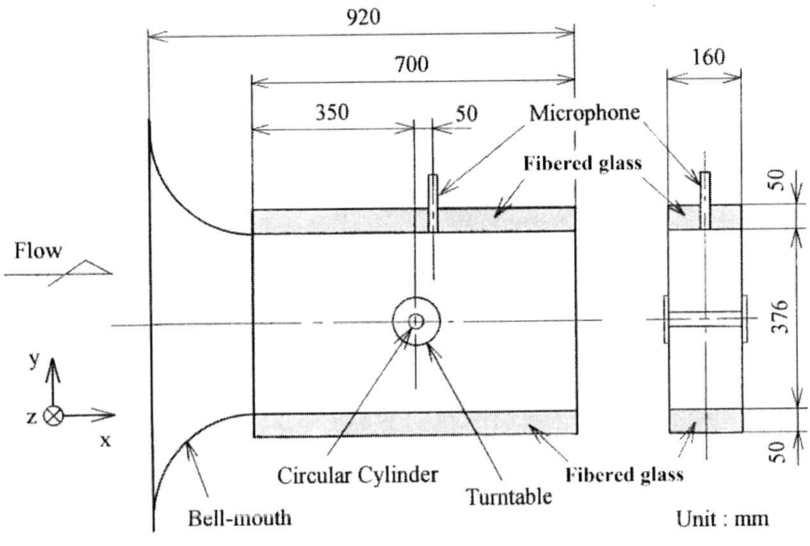

FIGURE 3. The schematic diagram of the test section (measurement section)

2.3. EXPERIMENTAL METHOD AND PROCEDURE

In advance of measurement of the fluid-dynamic noise, the flow velocity distribution in the test section is measured by a hot-wire anemometer, and the state of flow is understood. The relationship between the air flow velocity which passes the test section and the static pressure on the surface of wall is previously authorized using a

Pitot tube and a digital pressure gauge. Proofreading of a microphone and a precision noise level meter is performed using the piston phone (RION, NC-72, 250Hz, 114dB). The measurement procedure for the sound of flow is as follows. The test air flow velocity is set by operating the rotational speed controller of the blower. The fluid-dynamic noise is measured by the microphone, and the overall noise level and frequency analyses are done using the precision sound level meter and the fast Fourier transform analyzer. The flow velocity distribution in the measurement section and the measurement of the disturbance intensity relative to the main flow is as follows. The I type probe of the hot-wire anemometer is inserted detaching the microphone, it traverses in a vertical direction (y direction) at 5mm intervals (the interval of traverse is 2.5mm near the wall), and the air flow velocity is measured at the microphone installation position. The frequency of the oscillating flow due to Karman vortex shedding from the circular cylinder is measured as follows. The I type probe of the hot-wire anemometer is fixed in a position such that a clear shape of the waves can be obtained, and the output signal and frequency are using the fast Fourier transform analyzer. Here, averaging is performed ten times in the frequency analysis.

3. EXPERIMENTAL RESULT AND DISCUSSION

This chapter describes the result of having investigated about the basic characteristic of a producing wind tunnel, and the result of having performed sound verification.

3.1. THE FLUID-DYNAMIC CHARACTERISTIC AND THE ACOUSTIC CHARACTERISTIC OF A PRODUCING WIND TUNNEL

In order to understand the performance of a producing wind tunnel, investigation of the minimum flow velocity and the maximum flow velocity was performed using the Pitot tube. The minimum flow velocity in the test section was 2.5m/s, when the number of rotations of a fan was $100min^{-1}$, and the maximum flow velocity in the test section was 35m/s when the number of rotations of a fan was $1300min^{-1}$.

In a low noise wind tunnel, it becomes important especially to suppress propagation of the operation noise of the fan. Since this wind tunnel is a suction type wind tunnel, it is necessary to make it not accept fan generating noise in a test section. Accordingly, it is important not to leak the operation sound of the fan outside a fan room. So, the noise characteristic of the around of a wind tunnel was investigated. In order to understand the quietness of the wind tunnel, the sound pressure level around the test wind tunnel when it is driven or stopped was measured. Generally, the noise when the wind tunnel is operated is divided into air flow noise, and the operating noise of the blower. It is especially important in the fluid-dynamic noise measurement to suppress the propagation of the operating noise of the blower. The wind tunnel should not accept the blower generation noise in the measurement section. It is important that the operating sound of the blower does not leak outside the fan room. It is necessary, therefore, to understand the noise characteristics around the wind tunnel. The microphone positions for the noise measurement around the wind tunnel are shown in Fig. 4. Microphones are set up outside the fan room at a height of 1m off the ground, at measurement points (A-K). At measurement points (L1, L2) in the blower room, microphones are set up at a height of 1m, and placed a 700mm away from the electric motor and the blower outlet. Figure 5 shows the noise measurements at each measurement point when the circular cylinder is not set up in the measurement section and when the wind tunnel is in operation. The noise levels around the wind tunnel, almost the same, but differ inside and outside of the fan room, and when flow velocity increases, the difference increased. The noise levels inside and outside the fan room were 26dB and 32dB, respectively, when the wind tunnel was not operating. The level of sound intensity is defined by $L=10\log_{10}I/I_0$ (dB) (I_0 is an intensity of the sound of the standard: 10^{-12} W/m^2). Here, when the level of intensity of a sound inside the fan room is defined as *LIN*, and the level of intensity of a sound outside the fan room is defined as *LOUT*, the ratio of the level of sound-intensity *LIN/LOUT* is given by $LIN/LOUT = 10^{(LIN-LOUT)/10}$. The air flow velocity range is 5-35m/s, so the ratio of the level of sound-intensity *LIN/LOUT* becomes 155 -6760. Therefore, it is clear that the noise in the blower room is intercepted.

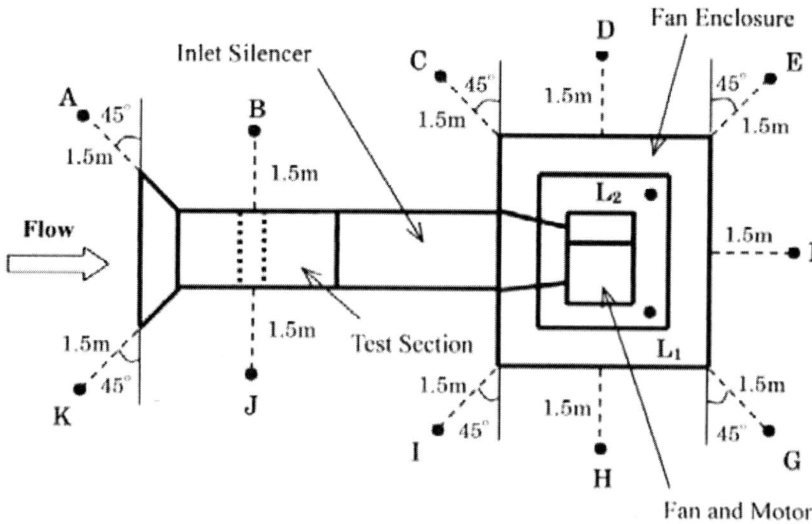

FIGURE 4. Sound measurement points around a wind tunnel pressure level

3.2. FLOW CHARACTERISTICS IN THE MEASUREMENT SECTION

Flow characteristics in the measurement section where the sound absorbing wall (fibered glass wall) had been used were investigated. The hot-wire probe was inserted in the microphone's installation position; it traversed in a vertical direction (y direction), and the velocity and disturbance intensity were measured. Figure 6 shows the velocity distribution and the disturbance intensity when the air flow velocity is U=28m/s. The velocity distribution (Δ symbol) for an

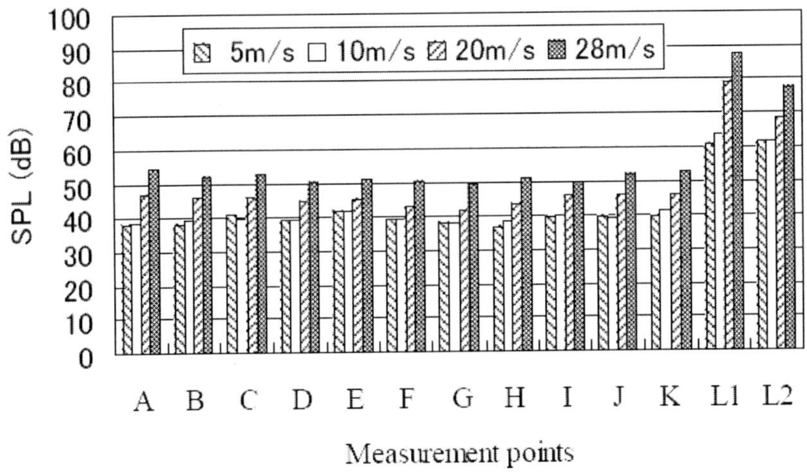

FIGURE 5. Sound pressure level around a wind tunnel with a flow velocity in the test section ranging from 5m/s to 28m/s

acrylic wall is plotted for comparison. Here, the abscissa is a measurement position. The center of width in the vertical direction of the measurement section is assumed to be zero points; the upper side is assumed to be + mark, and the lower side is assumed to be - mark. The measurement position is made dimensionless by width$B$=376mm in the vertical direction of the measurement section. The ordinate shows the velocity distribution and the disturbance intensity, respectively. The symmetry of a flow was good to the center and turbulence intensity was less than 0.5% in the range which is maintaining the equality of a flow. The disturbance intensity to keep the uniformity of the flow was within 0.5%. Here, if uniform flow velocity U=28m/s and the measurement point x=400mm are calculated by Prantl's exact solution ($\delta=0.22(\nu/Ux)^{0.167}x$), the thickness of the boundary layer is 9.6mm. The thickness of the boundary layer as in Fig. 6 is 10mm for the acrylic wall, and this agrees with the value from Prantl's expression. However, the thickness of the boundary layer above the sound absorbing wall was about 28mm, and was about three times that in Prantl's expression because the sound-absorbing wall was made of fibered glass with a rough surface. It was clarified that the wall was necessary to obtain a wide measurement section and thus improve the uniformity of the flow.

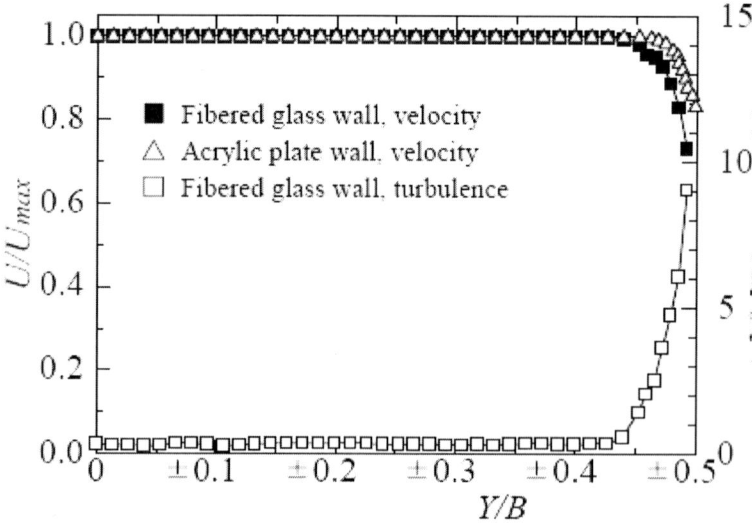

FIGURE 6. Flow velocity distribution and turbulence intensity in the test (measurement) section at a main flow velocity of U=28m/s

3.3. THE RELATION BETWEEN SOUND SOURCE AND MEASUREMENT POSITION

It is known that the fluid-dynamic noise made from the circular cylinder placed into the air flow is a dipole sound. Since there is single directivity also in a microphone, it is important to understand the influence on the measurement result by the spatial relationship of a sound source and its microphone.Figure 7 shows the measurement result of the sound pressure level when varying the distance x between centers of a circular cylinder and a microphone in the range from 5mm to 95mm. Here, the airflow velocity in a test section was U=28m/s and the diameter of circular cylinder was 20mm. It is understood that the measured sound pressure level is almost the same. So, in measurement of acoustic frequency, distance between centers of circular cylinder and microphone was set to 50mm. Here, it is expected that the pressure fluctuation of a short-distance field is included in the sound pressure which will have been measured if the measurement position of sound is generally near from a sound source (circular cylinder). In this study, the distance between the circular cylinder and the microphone was narrower compared with the device arrangement for an ordinary sound measurement. Because the noise measurement of the flow in the fluid machine such as the gas turbines and jet engines is assumed, and the measurement of the fluid-dynamic sound caused by the flow around the object such as the supports and umbones installed in the pipeline and the duct is assumed, it becomes such arrangement. Therefore, the influence of the near field appears to be strong, making a quantitative evaluation of the sound pressure level more difficult. Resolving this is a clear challenge for future studies. The relationship between the position rc by which the pressure fluctuation of a short-distance field can be disregarded now, and the minimum frequency f is given in $20\log(2\pi frc/a) >= 10$ dB (a is acoustic velocity) (Iida, 1996). Distance rc between the circular cylinder and the microphone becomes 188mm-211mm because the range of center-to-center spacing x between the circular cylinder and the microphone in this experiment is 5mm-95mm. The obtained lower critical frequency f becomes 910Hz-812Hz. When the center-to-center spacing is assumed to be 50mm, distance rc between two points becomes 194.5mm. The lower critical frequency f at that time is 880Hz.

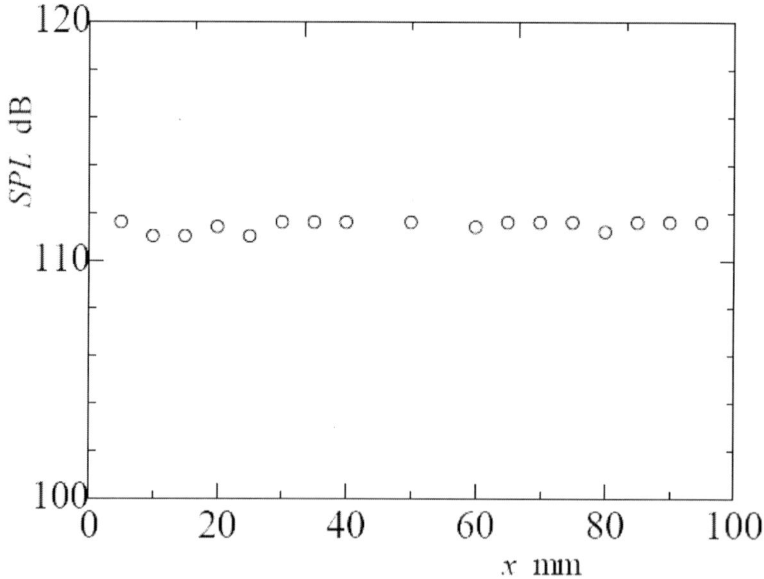

FIGURE 7. Measurement result for sound pressure level with a directivity check

3.4. MEASUREMENT RESULT AND VERIFICATION OF ACOUSTIC FREQUENCY

The back ground noise with acoustic half-free space of a test section was measured by making airflow velocity in a test section into U=28m/s. The circular cylinder of various diameters was installed in the test section, and frequency of a fluid-dynamic noise (acoustic frequency) was measured. Figure 8 shows the results of the acoustic frequency analysis with back ground noise (B.G.N.) in the measurement section and with a circular cylinder 20mm in diameter. The abscissa is frequency f, and the ordinate is the sound pressure level *SPL* in the figure. When the circular cylinder is set up in the measurement part, a peak at one big sound pressure level is obtained. At this time, the Strouhal number S $(= f\ d\ /U)$ calculated from the frequency f (=275Hz) and air flow velocity U (=28m/s) is S=0.2. It is considered that the microphone measures the acoustic frequency from the fluid oscillation based on Karman vortex shedding. The frequency of the oscillating flow behind the circular cylinder was measured using the hot wire anemometer for verification. Figure 9 shows the result of the frequency analysis using the microphone and the hot wire anemometer. The abscissa is frequency *f*, and the ordinate is a sound pressure level made dimensionless by the maximum value. In both measurement results, it is understood that one big peak is seen at the same frequency. Therefore, it was established that the acoustic frequency measured by the microphone was a fluid oscillating frequency based on Karman vortex shedding from the circular cylinder. This means that the fluid sound measured by making the acoustical free space can be measured in an internal flow. And, this measurement technique is considered suitable for the measurement of the fluid sound of an internal flow.

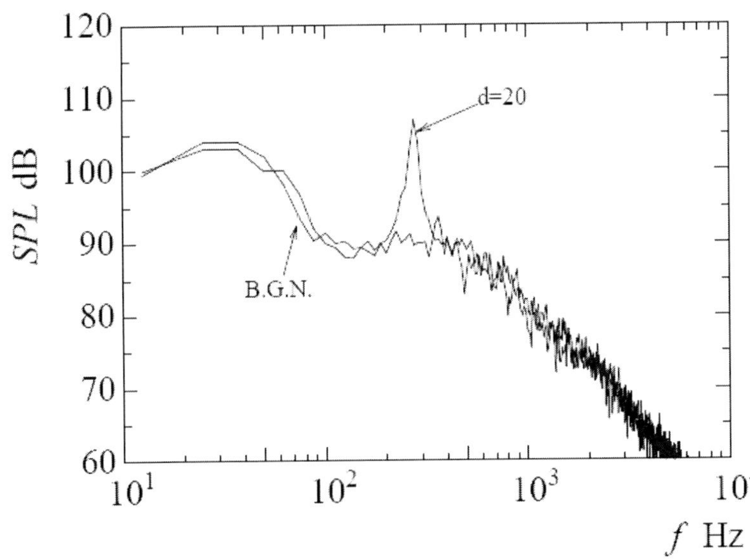

FIGURE 8. The frequency analysis of flow noise, U=28m/s, d=20mm

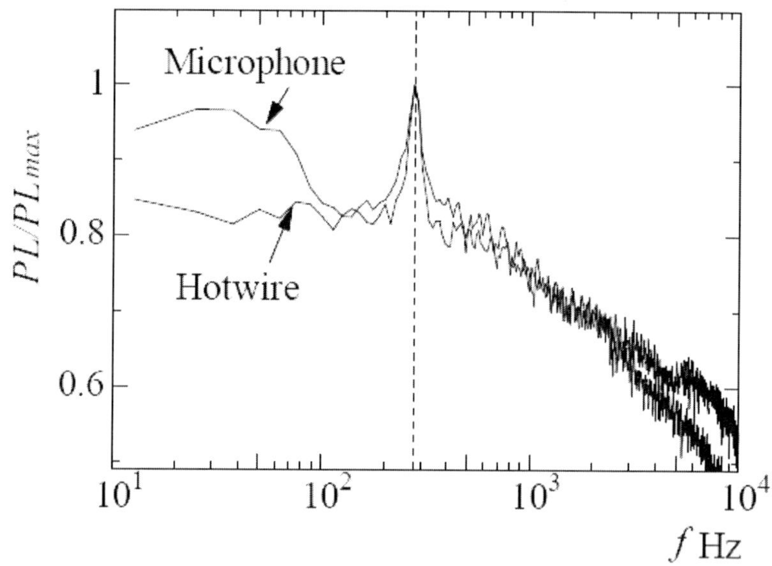

FIGURE 9. The relationship between the acoustic frequency and the fluid frequency due to vortex shedding

3.5. COMPARISON OF MEASUREMENT RESULTS WITH A BLOW-TYPE WIND TUNNEL

Figure 10 shows the variation of the peak frequency of the sound pressure level at the time of varying a circular cylinder diameter. The back ground noise is also shown for comparison. The abscissa is frequency f and the ordinate is a sound pressure level SPL. Increase of a cylinder diameter can see the tendency for a sound pressure level to increase and for peak frequency to decrease. The experimental result (Tomita et al., 1982) in the wind tunnel of a blow type with a half-opening type test section is shown in Fig. 11 for comparison with this experimental result. Although the variation of a sound pressure level or peak frequency to the variation of the diameter of the circular cylinder shows the same tendency, in each circular cylinder, one large peak and its harmonics component are seen, and spectrum distribution of the fluid-dynamic noise made when a circular cylinder is installed into an air current constitutes a larger sound pressure level than a back ground noise by the high frequency side which passed over the large peak. This suggests containing other sounds potential in not only the fluid-dynamic sound to be measured but also the flow noise. Therefore, it appears that the use of a blow-type wind tunnel with a half-open measurement section is rather inconvenient for measuring a sound effect. On the other hand, the results from a sealed-type measurement section of a suction-type wind tunnel becomes a sound pressure level that only the section of the frequency of the aimed fluid-dynamic sound is big as shown in Fig. 10, and the other frequency components are the same degree of the sound pressure level as the back ground noise. This is convenient for the examination of sound effects. The suction wind tunnel with a sealed-type measurement section can be expected to be a good measurement technique for examining sound effects.

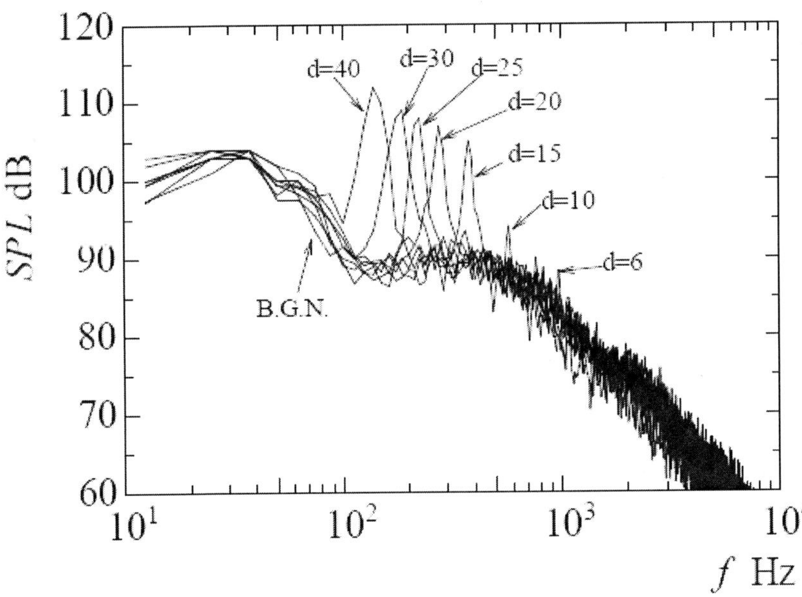

FIGURE 10. The characteristics of fluid-dynamic noise, in the case of present test section

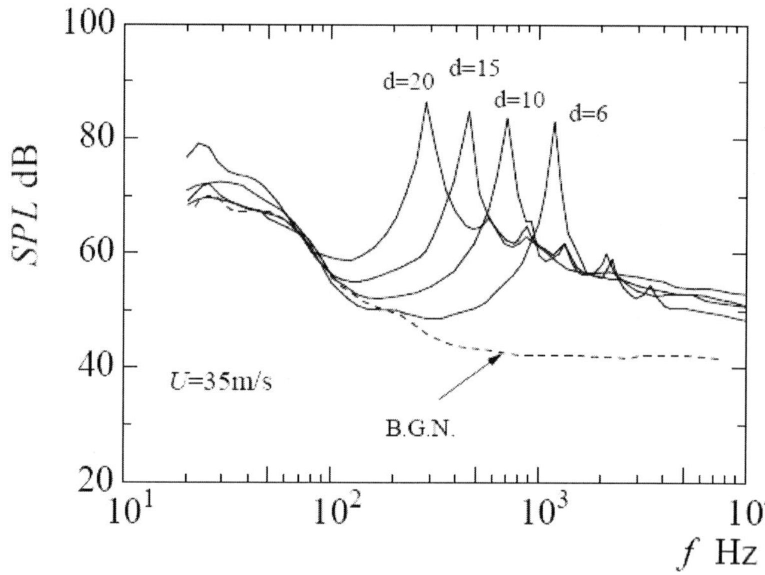

FIGURE 11. Characteristics of fluid-dynamic noise, in blow-type wind tunnel (Tomita et.al., 1982)

3.6. EFFECT OF ACOUSTIC MATERIAL AND SOUND DIRECTIVITY

In order to verify the effect of the sound-absorbing material (fibered glass) in the measurement section, the acoustic frequency from the circular cylinder was measured. At this time, the microphone is set up from the bell mouse to 500mm upstream side by equal height to the circular cylinder installation position. The air flow velocity of the measurement section was U=28m/s. As a result, the measurement result differed according to the existence of the sound-absorbing material. Figure 12 shows the measurement results for circular cylinders 20mm, 25mm, 30mm, and 40mm in diameter when upper and lower sidewalls made of an acrylic board are used. Here, the back-ground noise (B.G.N.) is shown for comparison. The sound pressure levels for the peak frequency of each circular cylinder are small, and the peak frequency is twice the value of fluid oscillating frequency, based on the Karman vortex shedding shown in Fig. 10. On the other hand, the measurement result when sound-absorbing material is installed is shown in Fig. 13, relative to back-ground noise (B.G.N.) at circular cylinder diameters of 20mm-40mm. Two sound pressure peaks are seen in each figure. The first peak (1st peak) on the low frequency side is a Karman vortex shedding frequency, and the second peak (2nd peak) on the high frequency side is twice the Karman vortex shedding frequency. Moreover, the magnitude correlation of the two peaks is different in each circular cylinder. In the case of circular cylinders with diameters of 10mm, 15mm, and 30mm, the first peak (1st peak) on the low frequency side is larger. In the case circular cylinder diameters of 20mm, 25mm, and 40mm, the second peak (2nd peak) on the high frequency side is larger. It is shown that there is a change in the interference pattern of the sound wave in a vertical direction in the flow in the measurement section. The two peaks can also be observed in a blow-type wind tunnel with a half-open measurement section as shown in Fig. 11. In this case, however, the microphone is set up at and angle of 45 degrees and positioned 500mm behind the circular cylinder, aiming at the sound around the circular cylinder. The first peak (1st peak) on the low frequency side is always larger than the second peak (2nd peak) on the high frequency side in the magnitude correlation of the peak because of the position of the microphone and the directivity of the microphone. A comparison of the results between a suction-type and a blow-type wind tunnel with sound-absorbing material installed shows that the acoustical free space of an internal flow can become an acoustical free space similar to the case of an external flow. The sound is fluctuation of the pressure which transmits the inside of fluid, the size of the amplitude is the size of sound, and the height of oscillation frequency is the height of sound. The fluid force acts on the circular cylinder by the fluid fluctuation according to the vortex shedding from the circular cylinder. The oscillation of the fluid force can be divided into a lift component and a drag component, at a ratio of 1:2. The circular cylinder placed on the air flow is a source of two kinds of sound waves as two peaks are apparent in the fluid sound. When acrylic upper and lower sidewalls are used, the acoustical free space becomes the only flow direction. The sound by the oscillation in the direction of the lift is canceled by acoustical interference. Therefore, the peak frequencies of each circular cylinder shown in Fig. 12 are considered to be the oscillation a sound from the drag direction. On the other hand, when the sound-absorbing material is installed on the upper and lower sidewalls, the acoustical free space is two directional (a parallel direction and a vertical direction) for the flow. It is an acoustical free space similar to the blow-type wind tunnel with a half-open measurement section. Therefore, the sounds of the lift and drag oscillations are measured as shown inFig. 13.

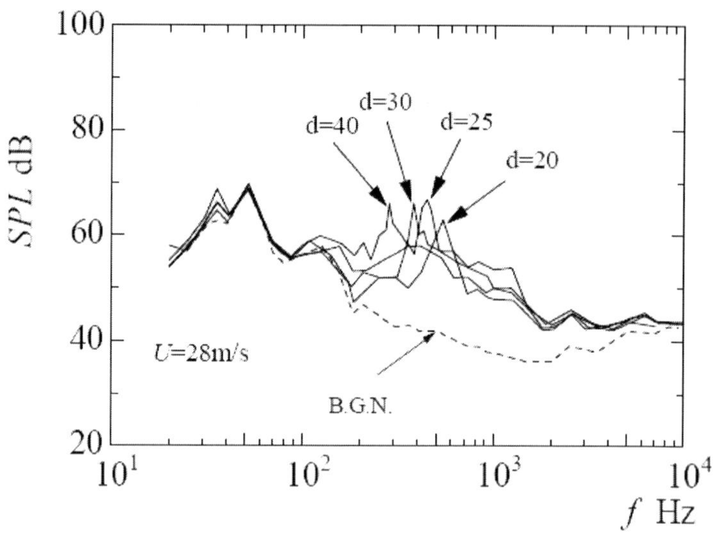

FIGURE 12. Characteristics of fluid-dynamic noise at the measurement point from the up-stream side of the test section, in the case of solid wall

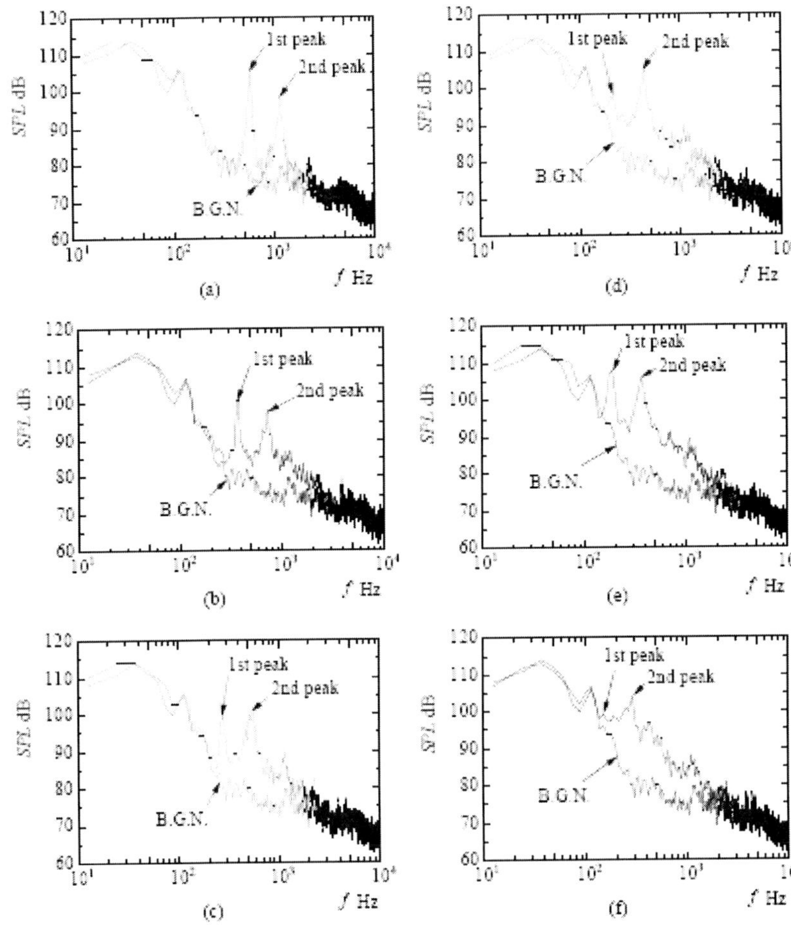

FIGURE 13. Characteristics of fluid-dynamic noise at the measurement point from the up-stream side of the test section, in the case of fibered glass wall; (a) cylinder diameter *d* is 10mm, (b) *d*=15mm, (c) *d*=20mm, (d) *d*=25mm, (e) *d*=30mm, (f) *d*=40mm

In general, because the oscillation amplitude of the lift is far larger than that of the drag, it is expected that the sound pressure level in the direction of the lift is far larger than the sound pressure of the drag direction. However, the sound from the oscillation of the drag direction is easily detected because the directivity microphone is located on the upstream side of the circular cylinder in this measurement, and the fluid-dynamic sound by the oscillation in the direction of the lift is not detected easily. In addition, because the interference pattern of the sound wave in the direction of the lift is different in each circular cylinder, the sound pressure level in the direction of the lift is small. The sound pressure level of the drag direction has a large value, as shown in Figs. 12(c) (d) and (f). Moreover, when the microphone is set up in the measurement section, only the fluid-dynamic sound in the direction of the lift is measured as shown in Fig. 8. Such a phenomenon suggests that the directivity of the sound source and the directivity of the microphone are at issue, and this is the subject of future investigation.

4. CONCLUSION

This study proposed a technique to measure the fluid-dynamic noise of an internal flow in a wind tunnel, and the fluid-dynamic noise from a circular cylinder placed on the air flow of a suction-type wind tunnel with a sealed-type measurement section with sound-absorbing material (fibered grass) was measured. The following conclusions were obtained.

1. The acoustic performance and fluid-dynamic performance of a test wind tunnel were good. The following results were obtained for the performance of the test wind tunnel. The noise in the blower room is effectively intercepted. The position of the sound source and the microphone are not influenced by directivity. The uniformity of the flow of the measurement section narrows when sound-absorbing material is used for the measurement section of the test wind tunnel.

2. The following results were obtained from installing sound-absorbing material in the measurement section. The acoustical free space can be made from the closed space.

When the surface of the microphone was arranged and set up on the surface of the sound-absorbing material, the measurement of the fluid sound of an internal flow became possible without any disarrangement of the flow-field.

3. The acoustic frequency measured by the microphone was confirmed to have a frequency based on the fluid oscillation caused by the Karman vortex shedding measured with the hot-wire anemometer.

4. The following results were obtained when a comparison was made with the results from a blow-type wind tunnel. The aimed acoustic frequency was measured by the large sound pressure level. Other frequency elements were the same degrees of the sound pressure level as the back ground noise. It has been understood that such a result was convenient when a sound effect was examined.

5. When an acoustical effect was examined, it was understood that the following consideration is necessary. The distance between the sound source and the microphone must be set in consideration of the influence of the pressure fluctuation of the near-field. The lower bound frequency must be understood. The microphone must be arranged in consideration of the sound directivity with the sound source.

6. From the results outlined in (2)-(4), this present measurement technique is considered to be a technique useful for the measurement of the fluid sound of an internal flow.

Nomenclature

a: acoustic velocity, m/s

B: height of test section, m

B.G.N.: back ground noise, dB

d: diameter of circular cylinder, m

f: frequency, Hz

I: intensity of sound, W/m^2

I_0: intensity of sound of standard, W/m^2

L: level of sound intensity, dB

L_{IN}: level of sound intensity inside fan room, dB

L_{OUT}: level of sound intensity outside fan room, dB

PL/PL_{max}: dimensionless sound pressure level, non-dimensional

r_c: critical distance, m

S: Strouhal number, non-dimensional

SPL: sound pressure level, dB

U: main flow velocity, m/s

U_{max}: maximum flow velocity, m/s

u: component of fluctuating flow velocity, m/s

X: measurement position in test section, m

x: coordinate component (flow direction), m

Y: measurement position in test section, m

y: coordinate component (vertical direction), m

z: coordinate component (horizontal direction), m

δ: thickness of boundary layer, m

ν: kinematic viscosity of air, m^2/s

REFERENCES

1. Fujita, H. (1994). The Present Condition and View for the Basic-study About the Flow and Sound Controls. *Proceedings of the 72nd the JSME Fall Meeting of the Japan Society of Mechanical Engineers,* pp. 360-363, Tokyo, Japan, October, 1994

2. Fujita, H. (1996). Experimental Study on Aerodynamic Noise Generated from two-Dimensional Models (1st Report, Study on Wind Tunnel Wall Effect and Wall Materials), *Transactions of the Japan Society of Mechanical Engineers, Series B,* Vol. 62, No. 593, (January 1996), pp. 187-193

3. Iida, A., Fujita, H., Kato, C., & Otaguro, Y. (1996). Experimental Investigation of the Generation Mechanism of Aerodynamic Noise (2nd Report, On Correlation Between Surface Pressure Fluctuation and Aerodynamic Sound Radiated from a Circular Cylinder), *Transactions of the Japan Society of Mechanical Engineers, Series B,* Vol. 62, No. 604, (December 1996), pp. 4160-4167

4. Mochizuki, O., & Maruta, Y. (1996). *Introduction of Fluid Sound Engineering*, Asakura-shoten, ISBN 4-254-23088-5, Tokyo, Japan

5. Tomita, S., Suzuki, S., Inagaki, S., Yokoyama, T., Kobayashi S., & Tsukamoto, G. (1982). The Study about Fluid Sound from a Cylinder in a Uniform Flow (1st Report, in the Case of Circular Cylinders), *Proceedings of the Thohoku Branchi Meeting of the Japan Society of Mechanical Engineers*, Senday, Japan, October, 1984

8

CHAPTER

Investigating Effect of Tunnel Gate Shapes with Similar Cross Section on Inserted Forces on Its Coverage and Soil Surface Settlement

Arash Rostami[1], Nima Asghari[2], Morteza Askari Ziarati[1], Saman Jahani[3], Bahman Shahi[4]

[1]*Department of Civil Engineering, Central Tehran Branch, Islamic Azad University, Tehran, Iran*

[2]*Department of Civil Engineering, Noshiravani University of Technology, Babol, Iran*

[3]*Department of Civil Engineering, UAE Branch, Islamic Azad University, Dubai, UAE*

[4]*Railway Engineering Department, Faculty of Civil Engineering, Isfihan University, Isfihan, Iran*

ABSTRACT

According to technology development and relative facilitation in digging and underground structures, ways, highways, all types of tunnels, underground train network, and other underground settle, storage are number of structure built and developed in advanced countries. In most situation, tunnel digging operations are done years after its construction or are not recorded in new structures regulations; therefore, this research investigates soil settlement and inserting force to tunnel coverage by limiting studies about effects of tunnel shapes on soil settlement using Plaxis, Seismo Signal, and Seismo Aspect. This study shows that rectangular tunnel has the most settlement in soil surface and circular tunnel has the least settlement but horseshoe tunnel has similar behavior to circular tunnel; however, earth subsidence level by digging this tunnel is more than circular tunnel. In addition, sectional shape has direct effect on inserting forces on tunnel coverage.

Keywords: Soil Settlement, Rectangular Tunnel, Circular Tunnel, Horseshoe Tunnel, Concrete Coverage

1. INTRODUCTION

Most buildings have been built based on earthquake standards before building each tunnel beneath it. On the other hand, sometimes reaction effects of structure-soil are considered in designing buildings but effects of tunnel digging are not considered. Yet, buildings designed and executed before considering tunnel digging with tunnel built under them definitely will be influenced by factors such as earth level movement and consequently structure movement. Another effective and important parameter in soil settlement is tunnel shape. Tunnels are usually built in rectangular, circular, and horseshoe shapes. Studies in this field include works of authors such as Park et al. (2009) [1] , Hue et al., (2006) [2] , Sanchez et al., (2009) [3] that each one individually evaluates effect of seismic loads of earthquake made on tunnel coverage. In addition, Wang et al., (2012) [4] state that tunnels above settlement have unpenetrated coverage more than tunnels with penetrated coverage. Heydari et al. (2003) [5] investigate ground level slope, external load, and cross section of tunnel on settlement by tunnel digging using Plaxis software and state that ground level slope doesn't influence on ground level settlement by tunnel digging. Moreover, section shape change and external load increase ground level settlement. Mohammad Ali Nezhad and Pakbaz (2011) [6] investigate effect of seismic effect on shallow rectangular and circular tunnels and state that seismic effect on rectangular tunnel is more than circular tunnels. Nikzad and Taebi (2011) [7] investigate surficial settlement control originated from tunneling in urbans using injection, congestion, jet, irrigation, and freezing methods and state that injection is the best method of optimizing soil and controlling soil settlement.

Mirhabibi and Sorush (2012) [8] investigate case study of twice tunnel effects in Shiraz by adjacent structures. He investigate tunnel-structure reaction by considering structure, height of tunnel digging, distance of two tunnels and claim that structure increases vertical settlement. Increase in structure width reduces earth vertical settlement level. Azadi et al. (2013) [9] by investigating reaction of tunnel and adjacent structure conclude that increase in tunnel distance reduces structure settlement and increase in tunnel digging height increases structure settlement. Musivand and AskariZiyarati (2014) [10] investigate earth level settlement of tunnel digging by inserting harmonic loads with various frequencies and state that right and left side of settlement curve changes from sloppy to horizontal state by increase in frequency. Rostami et al. (2016) [11] investigate the effect of tunnel maintenance on surface settlement. Rostami et al. [11] (2016) evaluated the effect of tunnel maintenances on surface and concluded that an umbrella method had significant effect on surface. Rostami et al. [12] (2016) evaluated effect of tunnels with different shapes on ground acceleration and conducted that the rectangular tunnel have a significant effect on ground responses. Alielahi et al. (2014) studied the use of boundary element method on time-domain for seismic response of buried cylindrical holes in soil. As a result of this research; the numerical results suggested that in a random SV wave with buried hole, horizontal component at the top of the cavity is reduced in comparison with displacement of the empty earth [13] . Also Alielahi et al. [14] (2015) worked on seismic ground amplification by unlined tunnels and concluded that The effect of the tunnel on the seismic ground surface response is gradually decreased or becomes insignificant with increasing the buried depth of the tunnel.

2. METHODOLOGY

In present research, 2 types of soil with recorded specifications in Table 1 have been used so that it can be claimed that soil type (2) is harder than soil type (1). Each soil has a layer and don't have underground water.

Table 1. Soil specifications (Afifpour et al., 2011)

Row	Model No.	Wave Velocity (m/s)	Height (m)	Length (m)	Behavioral Model Type	Saturated Specific Weight (KN/m³)	Dry Specific Weight (KN/m³)
1	1	105.3	50	200	Mohr-Coulomb	17	17
2	2	290.3	50	200	Mohr-Coulomb	17	17

Modulus of Elasticity (KN/m²)	Friction Angle (degree)	Dilation Angle (degree)	Cohesion (KN/m²)	Poisson's Ratio	Mutual Resistance Coefficient	Rayleigh Alpha	Rayleigh Beta
50,000	29	5	0.4	0.3	0.7	0.01	0.001
380,000	29	5	10	0.3	0.7	0.01	0.001

The mentioned soil is homogeneous and has elastic-perfectly plastic behavior with Mohr-Coulomb criterion and includes of 5 elastic module equations such as internal friction ϕ, dilation angle ψ, CohesionC, soil Poisson's ratio v, behavioral model equations. Shear modulus G can shear wave velocity VS can be calculated based on relations using the mentioned parameters and soil specific weight γ. In addition, interface resistance coefficient has been considered 0.7.

3. TUNNEL SPECIFICATIONS

Tunnels have been dug in circular in various height placement, horseshoe, and rectangular forms by similar cross sections and constant height placement were dug in soil and concrete coverage was used for tunnel wall based on Table 2. Tunnel wall was modelled using flexural element which has elastic behavior. In addition, axial rigidity (EA) and flexural rigidity (EI) and coverage weight (w) can be obtained by coverage thickness (d) and concrete specific weight (γ).

Soil layer and tunnel were modelled by Plaxis software. In mentioned software, the mentioned soils in Table 1 were modelled then tunnels with different cross sections and height will be dug. Tunnel concrete coverage and structure foundation were also modelled by flexural element and elastic behavior and materials parameters are attributed to them. In order to define interface of elements (such as structure mutual foundation and adjacent soil, etc.), interface element was used. Meshing is the next step in software, then model was dynamically analyzed by inserting side forces as harmonic sine (with different frequencies) and seismic real records. Then needed responses such as seismic acceleration record in soil level will be extracted.

Tunnel Modelling

Figure 1 shows the general scheme of model. In the mentioned figure, d is tunnel diameter, h is placement height of tunnel (distance of tunnel enter from ground level) and b is structure foundation width. The dug tunnels in model is regarding to the following tables. In Table 3, tunnel shape and placement height is constant but tunnel diameter is varied. In this state, the ratio of diameter to height (d/h) will be considered varied. In Table 4, tunnel shape and diameter is constant and height placement is varied, so the ratio of h/d is varied in this table. In Table 5, tunnel shape is varied and other parameters are constant. The aim to say height placement is distance of tunnel center from ground level.

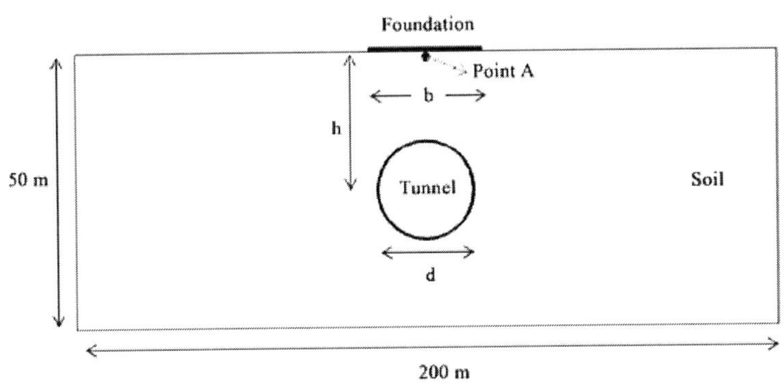

Figure 1. General scheme of tunnel.

Table 2. Tunnel coverage specifications (Maleki et al., 2011).

Behavior Type	Thickness (m)	Axial Rigidity (KN/m)	Flexural Rigidity (KN/m^3)	Specific Weight (KN/m^3)	Weight (KN/m^2)	Poisson's Ratio	Rayleigh Alpha	Rayleigh Beta
Elastic	0.35	8,050,000	82,180	2400	8.4	0.25	0.01	0.001

Table 3. Specification of circular tunnel with different diameters.

Row	Tunnel Shape	Height Placement	Diameter	Ratio of d/h
1	Circular	25	8	0.32
2			12	0.48
3			16	0.64
4			20	0.8

Table 4. Tunnel specifications with different heights.

Row	Tunnel Shape	Diameter	Height Placement	Ration of h/d
1	Circular	16	12	0.75
2			18	1.125
3			24	1.5
4			30	1.875
5			36	2.25
6			42	2.625

Table 5. Tunnel specifications with different shapes.

Row	Tunnel Shape	Diameter (m)	Height Placement (m)
1	Circular	16	25
2	Horseshoe		
3	Rectangular		

To define boundary conditions, constant and energy absorber boundaries are used. In constant boundary, roller support is considered for vertical lines, joint support is considered for horizontal line of soil and horizontal line above soil is without any support. In this way, it has movement in vertical direction and don't have any movement in horizontal direction. While inserting dynamic loads, waves reflect on model boundaries for turbulence. To prevent from intensive reflections, energy absorber boundaries in low part, left, and right sides are used. Analysis software meshes model to 6 or

15 dots triangles that are called meshing. In Plaxis software, average meshing is used. In addition, more sensitive areas meshing such as smaller tunnel walls are considered. Inserted records in this research are harmonic sine and real earthquake accelerograms. Table 6 shows sine record with different frequencies. In the mentioned table, p is record range and its value is 0.2. Moreover, the relationship between rotational frequency (ω) and natural frequency (f) is as following. Record duration of each sine record is 30 second.

Figure 2 shows sine records with 0.2 range and 2 Hz frequency. In addition, real seismic records are shown in Figures 3-5. These records are all compared to maximum 0.2 gr acceleration in Figures 6-8.

$$\omega = 2\pi f \text{ Angular frequency } \omega$$

4. DYNAMIC ANALYSIS IN MODEL IN PLAXIS

In this research, analysis time was 30 second with 0.03 second interval. In addition, equations were solved by numerical methods of Newmark design.

5. DISCUSSION AND RESULTS

Investigating Vertical Settlement of Ground Level. Vertical settlement of ground level for soil type (1) and (2) are considered under sine records with 2 settlement frequencies under tunnel effects with different crosses (but with similar cross sections). Finally, obtained results in soils (1) and (2) will be compared with each other.

Tunnel cross shape is another variable which effects on soil vertical level settlement will be investigated. In this state, three circular, rectangular, and horseshoe crosses will be used with similar cross sections in constant height. Figure 8 shows cross section in frequency (2). In figures CT, HT, RT indicate tunnel with circular, rectangular, and horseshoe sections, respectively. As it is seen from figures, tunnels with rectangular section make more vertical settlement in ground level. Of course, its reason is directly related to tunnel sections. Circular tunnels make the least settlement.

Figure 9 shows effects of sections for soil type (2) in frequency of 2 Hz. Circular tunnel makes the least set tlement and rectangular tunnel makes the most settlement. On the other hand, settlement of soil type (2) is more than soil type (1) for being stiffer.

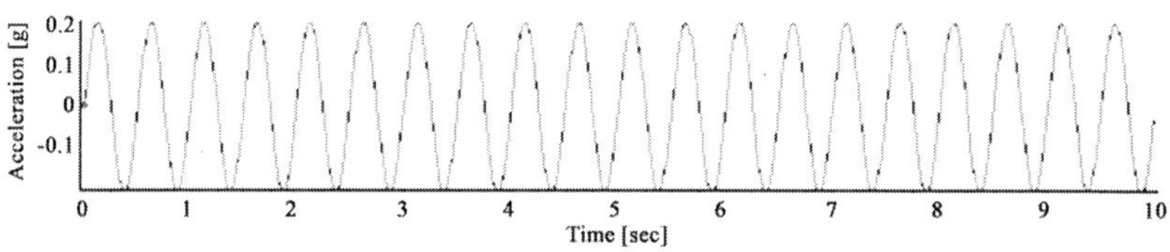

Figure 2. Sine record with 2 Hz frequency.

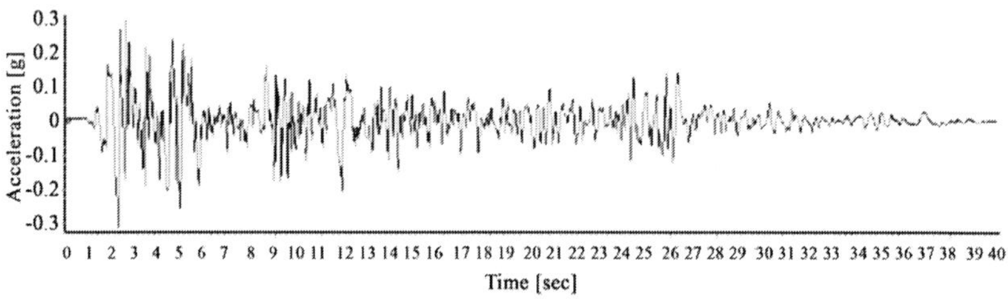

Figure 3. El Centro earthquake record.

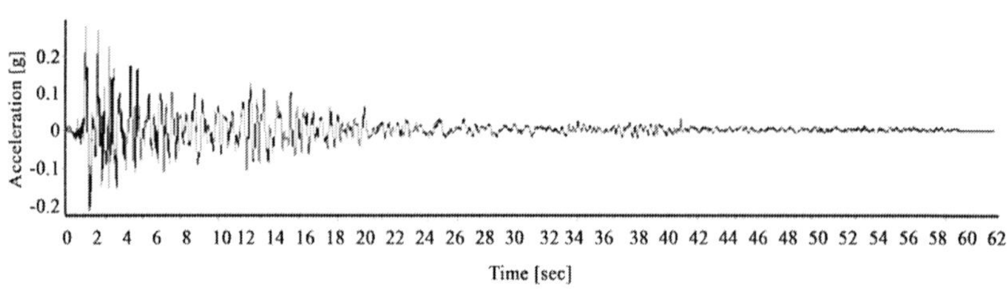

Figure 4. San Fernando earthquake record.

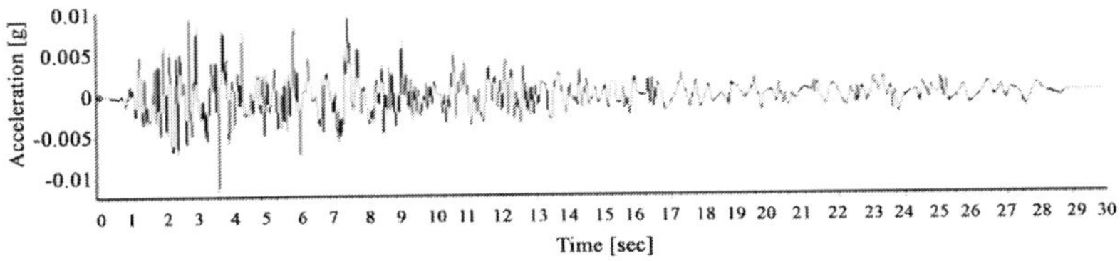

Figure 5. San Luis earthquake record.

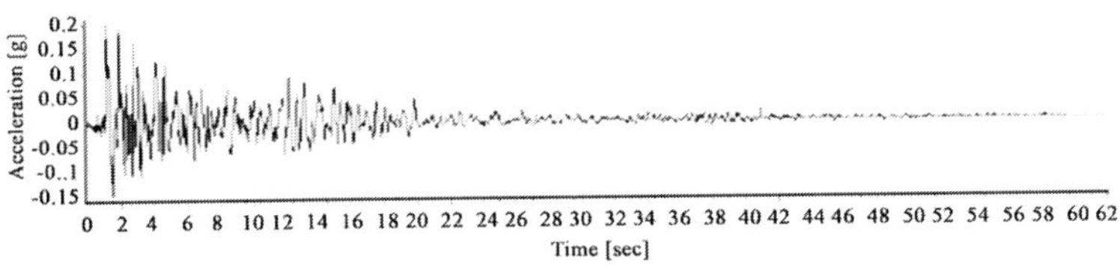

Figure 6. Record compared to 0.2 g of San Fernando earthquake.

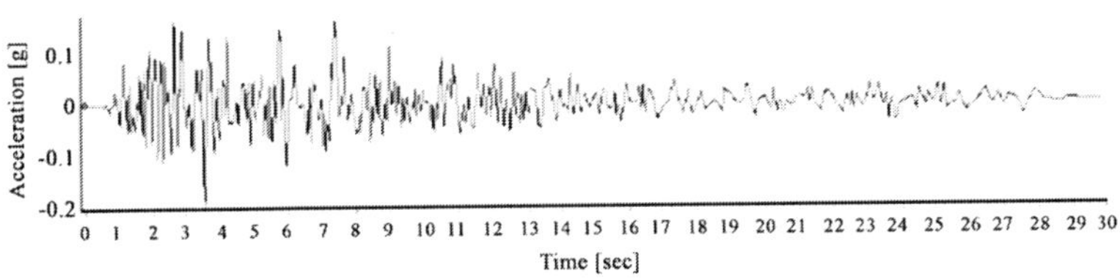

Figure 7. Record compared to 0.2 g of San Fernando earthquake.

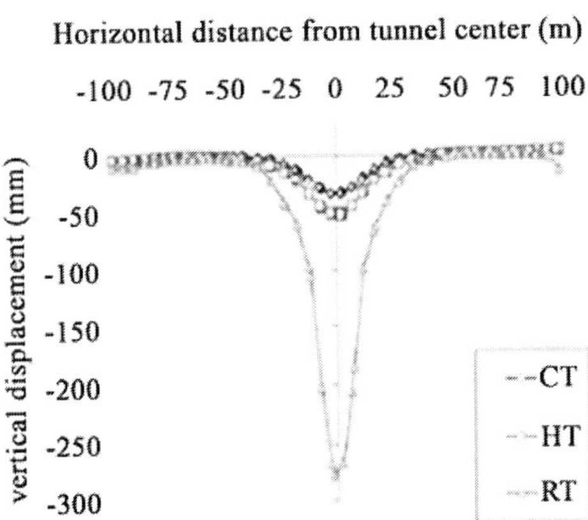

Figure 8. Vertical settlement of ground level in soil with 2 Hz frequency.

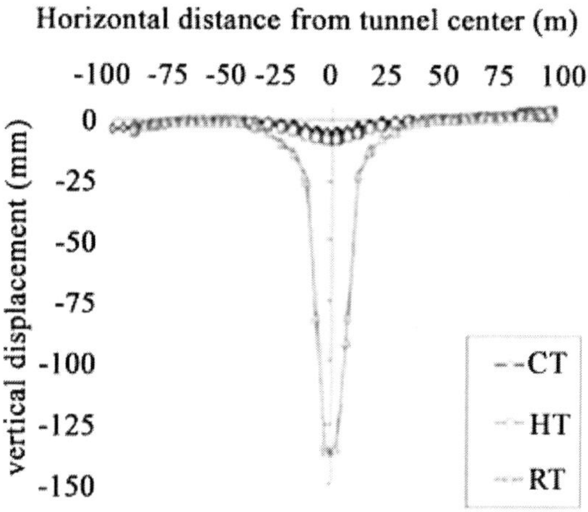

Figure 9. Vertical settlement of ground level in soil type (2) with 2 Hz frequency.

Table 6. Specifications of harmonic sine record.

Sine Record	Natural Frequencies (Hz)			
P sin (wt)	2	4	6	8

Therefore, sections have direct relation with soil level settlement. On the other hand, soil stiffness reduces movement but in both soil types, rectangular section has the most and circular section has the least settlement.

5.1. Inserted Forces on Tunnel Coverage

In this part, inserted forces on tunnel coverage (including axial force, flexural anchor, and shear force) were considered under sine records with 2Hz frequency for soil types 1 and 2 and tunnels with different sections (but with similar cross sections).

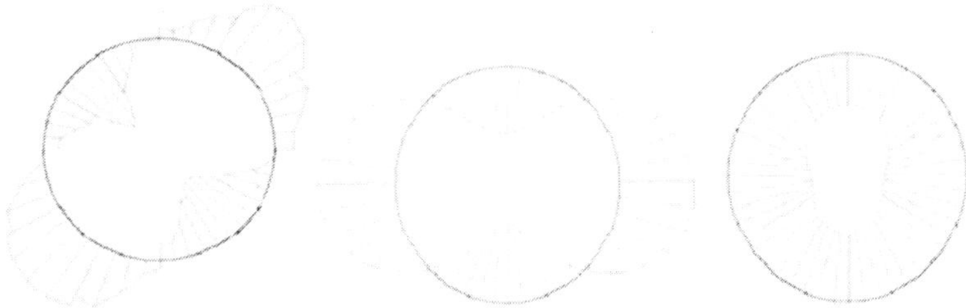

(a)

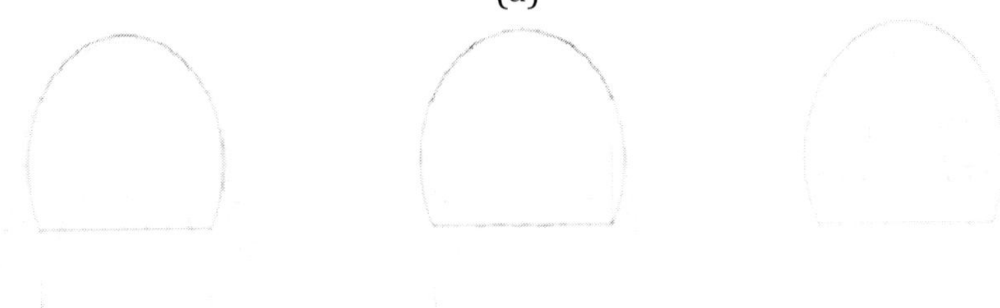

(b)

(c)

Figure 10. (a) General scheme of inserted forces on circular tunnel walls from right side includes axial force, flexural anchor, and shear force; (b) General scheme of inserted forces on horseshoe tunnel walls from right side includes axial force, flexural anchor, and shear force; (c) General scheme of inserted forces on rectangular tunnel walls from right side includes axial force, flexural anchor, and shear force.

5.2. Inserted Forces on Tunnel Coverage with Different Sections under Sine Records with (2) Hz Frequency in Soil Type (1)

Figures 10 shows general scheme of inserted forces on tunnel walls with circular, rectangular, and horseshoe sections with similar areas in constant height. It is concluded by examining these figures that forces distribution in more homogenous in circular tunnel than other tunnels. In addition, the most shear force and flexural anchor happen for horseshoe tunnel in its low part and for rectangular tunnels on the corners. Inserted forces values on tunnel walls are shown in Table 7 and Table 8. As it is seen in table, inserted forces on tunnel coverage in circular tunnels are less but the most value is made in rectangular tunnels.

Table 7. Inserted forces on tunnel walls with different cross sections in soil type (1) with 2 Hz frequency.

Row	Cross Section	Axial Force (KN/m)	Flexural Anchor (KN/m)	Shear Force (KN/m)
1	Circular	1860	80.71	23.5
2	Horseshoe	1460	807.16	855.77
3	Rectangular	1860	3440	1760

Table 8. Inserted force on tunnel walls with different cross sections in soil type (1) in (2) Hz.

Row	Cross Section	Axial Force (KN/m)	Flexural Anchor (KN/m)	Shear Force (KN/m)
1	Circular	1590	13.51	5.75
2	Horseshoe	1250	412.06	454.24
3	Rectangular	1990	2570	1580

Inserted forces on tunnel walls with different cross sections under sine records with 2 Hz frequency in soil type (2). Table 8 shows inserted forces on circular, horseshoe, and rectangular tunnel walls in soil type (2) with 2 Hz frequency. As it is seen. Inserted forces on circular tunnels coverage is less than other tunnels for different cross sections but rectangular tunnels make the most force.

5.3. Comparing Inserted Forces on Tunnel Coverage with Different Cross Sections under Sine Record with 2 Hz in Soil Types (1) and (2)

Cross sections have direct effect on inserted forces on tunnel coverage. Examining registered numbers in the mentioned table showed that inserted forces on circular tunnels coverage was less than other tunnels but rectangular tunnels made the most force in both soil types. Soil becoming stiff reduces forces. Moreover, it is concluded that forces distribution in circular tunnels is more homogeneous than other tunnels. The most value of shear force and flexural anchor happens for horseshoe tunnel in low part and for rectangular tunnel in corners.

5.4. Investigating Vertical Settlement of Ground Level

Figures 11-13 show vertical settlement of ground level by digging tunnels with different section in oil type (2). As it is indicated, the least ground level settlement was made by digging circular tunnels and the most made by digging rectangular tunnel. Moreover, for homogeneity, the maximum related acceleration to three records was 0.2 g settlement made by them in each relatively similar tunnel. Vertical settlement of ground level was shown in Figures 14-16 by digging tunnels with different sections in soil type (2). As it is seen, the most settlement is made by digging rectangular tunnel and the least settlement is made by circular tunnels. In addition, settlement made by them in each tunnel was relatively similar in both soil types for similarity of maximum related acceleration to three, 0.2 g records.

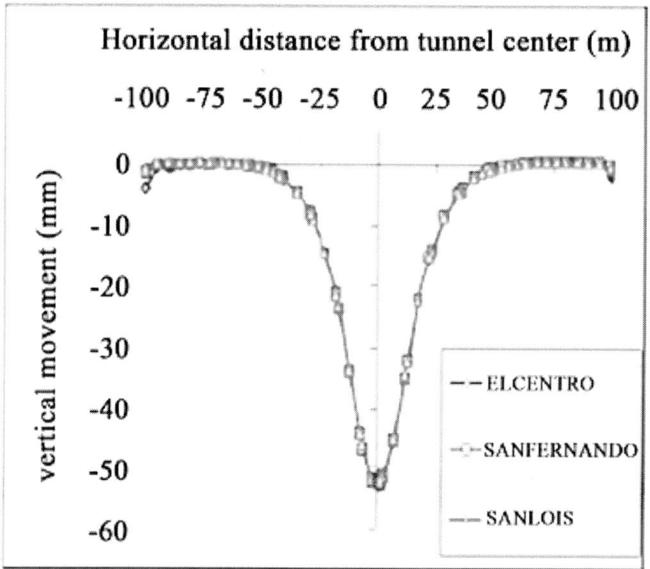

Figure 11. Vertical settlement of ground level in soil type (1) under circular tunnel.

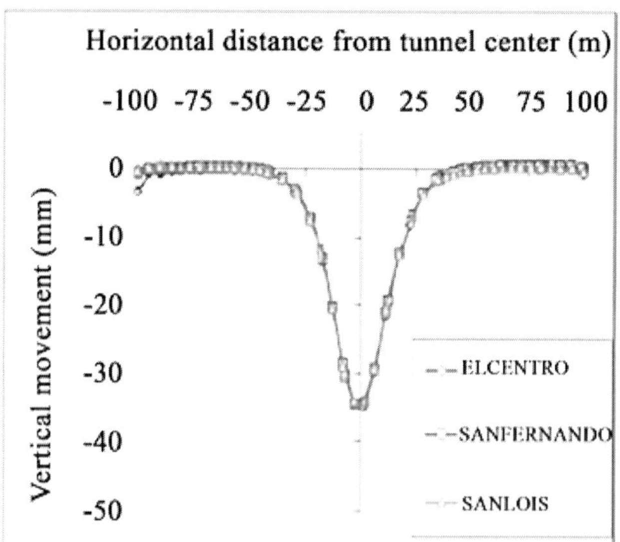

Figure 12. Vertical settlement of ground level in soil type (1) under horseshoe tunnel.

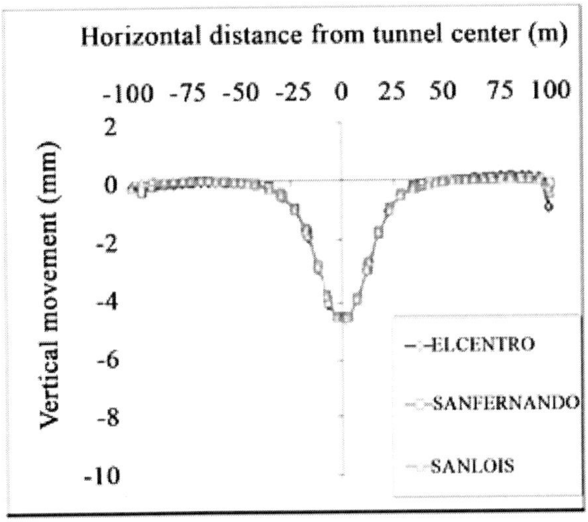

Figure 13. Vertical settlement of ground level in soil type (1) under rectangular tunnel.

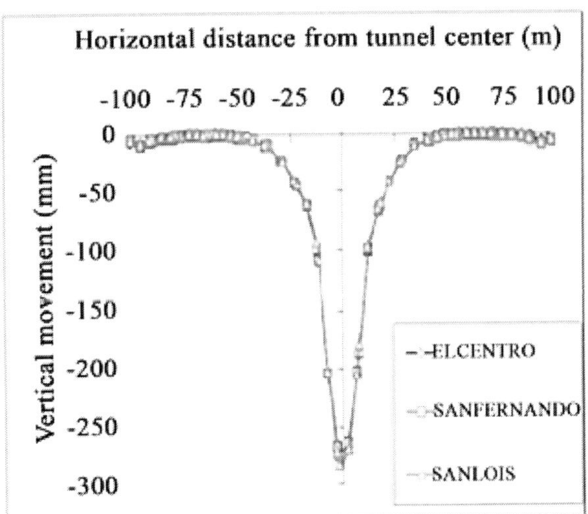

Figure 14. Vertical settlement of ground level in soil type (21) under circular tunnel.

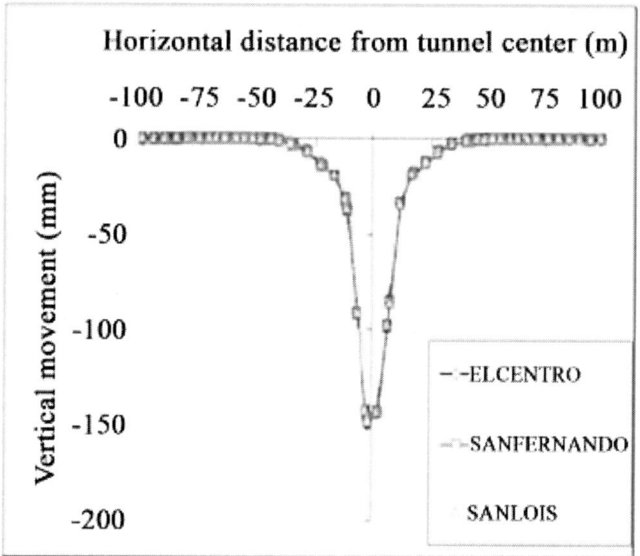

Figure 15. Vertical settlement of ground level in soil type (2) under horseshoe tunnel.

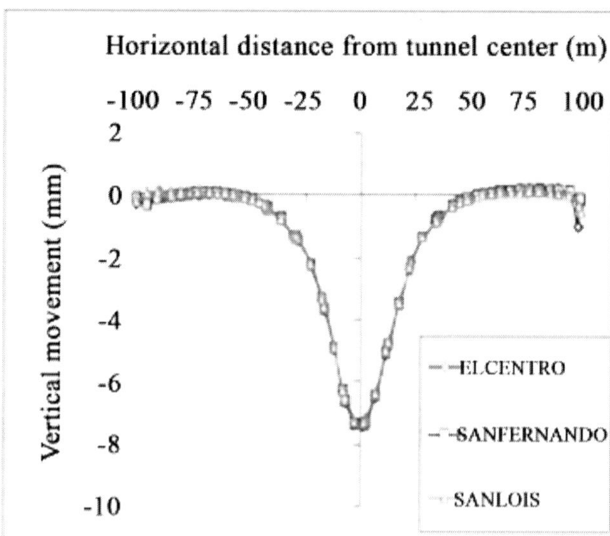

Figure 16. Vertical settlement of ground level in soil type (2) under rectangular tunnel.

5.5. Verification of Modelling

In order to verify model, obtained results of sine record with 2 Hz frequency was compared with San Fernando earthquake (with dominant frequency of 1.83 Hz) in soil type (1). It is expected to obtain similar results for nearing of 2 records frequencies. In this part, rectangular and circular tunnels are used based on Table 5. Figures 17-20 show vertical settlement of ground level by digging circular and rectangular tunnels under 2 Hz frequency and San Fernando earthquake. As it is observed, results of 2 different records are similar.

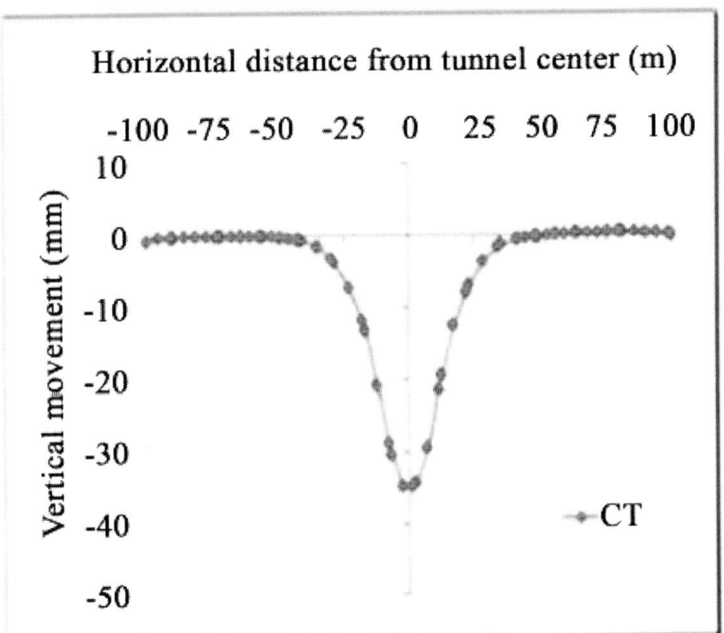

Figure 17. Vertical settlement of ground level by sine record of 2 Hz.

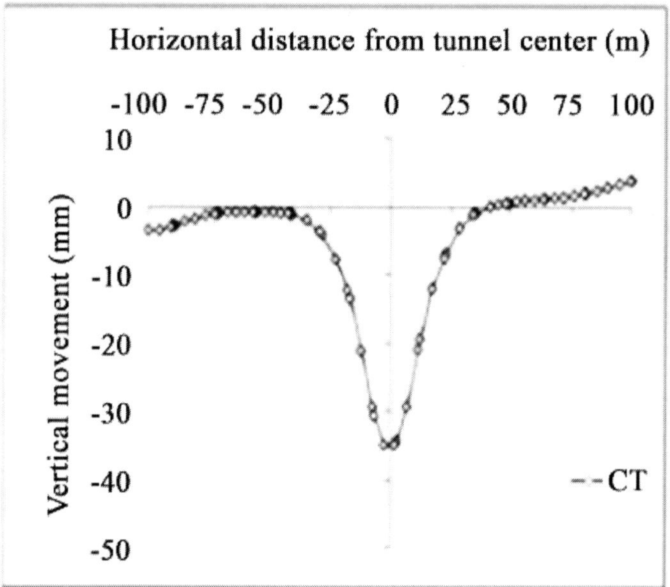

Figure 18. Vertical settlement of ground level by sine record of San Fernando.

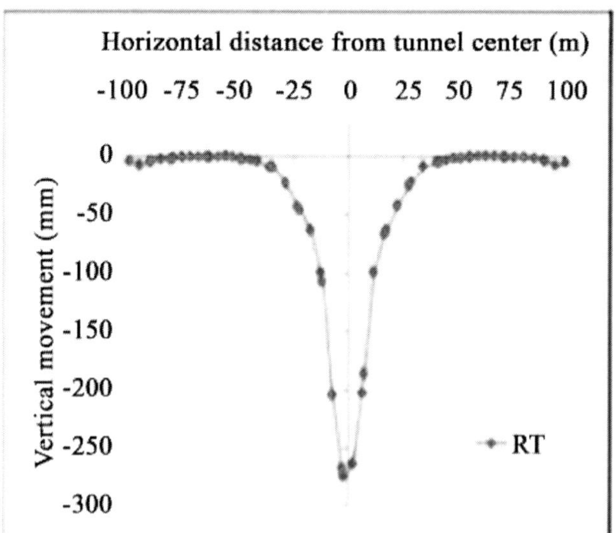

Figure 19. Vertical settlement of ground level by sine record of 2 Hz.

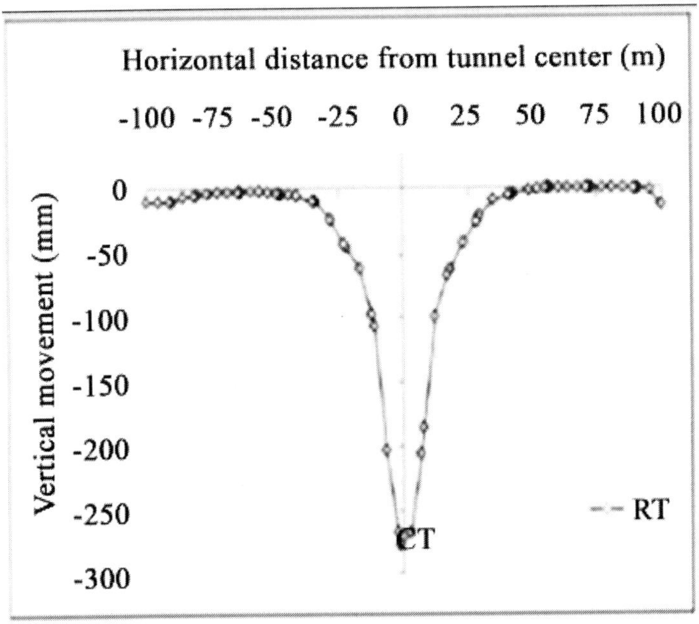

Figure 20. Vertical settleme t of ground level by sine record of San Fernando.

6. CONCLUSION

Digging tunnel in any diameter, placement height and shape make settlement in ground level. As much stiffer soil is, ground vertical settlement is less for soft soil. As much diameter or placement height of tunnel increases, vertical settlement of ground level increases. In tunnels with different cross sections shapes (with similar cross section), rectangular tunnel makes the most settlement in soil level, and circular tunnel makes the least settlement, but horseshoe tunnel has similar behavior to circular tunnel; however, ground level settlement made by digging this tunnel is more than what is made by digging circular tunnel and section shape has direct effect on inserted forces loads on tunnel coverage. Investigations show that inserted forces on circular tunnel walls coverage are less than others for its cross sections shape but rectangular tunnels insert the most forces and soil stiffness reduces forces. Moreover, it is concluded that forces distribution in circular tunnels is steadier than other tunnels. The most shear force and flexural anchor for horseshoe tunnel are in its low part and for rectangular tunnel is on its corners.

REFERENCES

1. Peck, R.B. (1969) Deep Excavation and Tunneling in Soft Ground. State-of-the-Art Report. Proceedings of the 7th International Conference on Soil Mechanics and Foundation Engineering, Mexico, 1969, 225-290.

2. Huo, H., Bobet, A., Fernandez, G. and Ramirez, J. (1996) Analytical Solution for Deep Rectangular Structures Subjected to Far Field Shear Stresses. Tunneling and underground space Technology, 21, 613-625.

3. Sánchez-Merino, A.L., Fernández-Sáez, J. and Navarro. C. (2009) Simplified Longitudinal Seismic Response of Tunnel Lining to Surface Wave. Soil Dynamics and Earthquake Engineering, 29, 579-582.

4. Sun, C. and Wang, Q. (2012) Effects of Underground Structure on Acceleration Response of Site. Advanced Materials Research, 368-373, 2791-2794.

5. Heydari, M.R., et al. (2003) Effect of Ground Slope, External Load, and Tunnel Cross-Sectional Shape of Earth Subsidence by Digging Soil Tunnels. 6th Conference of the Tunnel, College Civil, Iran Industry and Science University, Tehran, 27-29 January 2004.

6. Mohamamd, A., Pakbaz, R. and Sirus, M. (2011) Seismic Effect on Shallow Rectangular Tunnels. The 6th National Conference of Civil Engineering, 26-27 April 2011, Semnan University, Semnan.

7. Mahdi, N. and Vahid, T. (2011) Controlling Surficial Settlement in Tunneling in Urban Area Using Improvement

of Earth Specification. 1st National Conference of Zibakenar Civil and Development, Zibakenar, March 2014.

8. Mirhabibi, A. and Soroush, A. (2012) Effects of Surface Buildings on Twin Tunnelling-Induced Ground Settlements. Tunnelling and Underground Space Technology, 29, 40-51.

9. Azadi, M., Pourakbar, S. and Kashfi, A. (2013) Assessment of Optimum Settlement of Structure Adjacent Urban Tunnel by Using Neural Network Methods. Tunnelling and Underground Space Technology, 37, 1-9.

10. Mohsen, M. and Meysam, B. (2013) Investigating Effect of Earth Sloppy Effect on Ground Level Settlement by Digging Tunnel. 1th National Conference on Engineering and Technology, Aredil, 20 May-23 October 2013.

11. Rostami, A., Kamali-Dehkordi, P., Askari-Ziarati, M., Jahani, S. and Lotfi, K. (2016) The Types of Tunnel Maintenance in Umbrella Arch Method. Open Journal of Civil Engineering, 6, 156-162.

12. Rostami, A., Askari-Ziarati, M., Shahi, B. and Jahani, S. (2016) Evaluation of Seismic Behavior and Earth's Surface Acceleration, by Interaction of Tunnels with Different Shapes and Different Types of Soils. Open Journal of Civil Engineering, 6, 242-253.

13. Hamid, A. (2012) Applying a Time-Domain Boundary Element Method for Study of Seismic Ground Response in the Vicinity of Embedded Cylindrical Cavity. International Journal of Civil Engineering, 11.

14. Alielahi, H., Kamalian, M. and Adampira, M. (2015) Seismic Ground Amplification by Unlined Tunnels Subjected to Vertically Propagating SV and P Waves Using BEM. Soil Dynamics and Earthquake Engineering, 71, 63-79.

9

CHAPTER

Investigation of Drying Mechanism of Solids Using Wind Tunnel

Abdulaziz Almubarak[1]

[1] College of Technological Studies, Department of Chemical Engineering, Kuwait

1. INTRODUCTION

Drying of solids provides a technical challenge due to the presence of complex interactions between the simultaneous processes of heat and mass transfer, both on the surface and within the structure of the materials being dried. Internal moisture flow can occur by a complex mechanism depending on the structure of the solid body, moisture content, temperature and pressure in capillaries and pores. External conditions such as temperature, humidity, pressure, the flow velocity of the drying medium and the area of exposed surface also have a great effect on the mechanisms of drying.

Theoretical and experimental studies [1-6] reported the forced convection heat and mass transfer across different shapes. Most of these studies have considered a heated solid surface with a uniform surface temperature. However, this situation is not the same as in the drying process, where heat and mass transfer occurs simultaneously and the interfacial temperature and moisture content vary during the drying.

Evaporation from a flat plate surface to a laminar boundary layer was theoretically analyzed [7]. It was calculated distributions of the interfacial temperature and local Nusselt and Sherwood numbers for a parallel flow where both Prandtl and Schmidt numbers are unity. A conclusion stated that the characteristics of heat and mass transfer are highly conjugated and significantly influenced by the temperature dependency of vapor-liquid equilibrium, the magnitude of the latent heat of the phase change, and the thermal conductance of the flat plate. The work needs to be extended for the case of unsteady state conditions and to be repeated for a drying bed.

The variation in the surface temperature for a flat bed of a capillary porous material was discussed in [8]. The authors presented graphs that show an increase in surface temperature from the leading edge of the flat plate during the periods of drying. They also showed graphs in which the surface temperature decreased. In general, a physical meaning for that contradiction was needed to be considered.

As in [9], Jomaa *et al.* presented a simulation of the high-temperature drying of a paste in a scaled-up wind tunnel. They attempted to study the influence of the local air flow and thermal radiation on the drying behaviour of the product. A rapid air velocity (8 m s^{-1}) was used in the experiment, and an empirical model was derived to predict temperature and solvent content along the conveyer belt. Comparison of the experimental results with those predicted ones by the model showed a satisfactory agreement.

Evaporation of a pure liquid droplet has been widely studied both theoretically and experimentally [10-14]. However, in spray dryers and droplet drying applications, the droplets always consisted of multi-component mixtures of liquids and sometimes dissolved solids, forming a complicated multiphase composition. This makes analyzing heat and mass transfer processes more difficult. This effect is attributed to the presence of various components that vaporize at different rates, giving rise to a gradient in concentration in the liquid and vapor phase. In addition, a solid crust forms at the outer surface of droplet which acts as a resistance to heat and mass transfer processes.

Various experimental techniques have been used [15-19] to study the mechanisms of drying for a single droplet containing dissolved solids. The droplet was suspended freely from the end of a stable nozzle fixed in a wind tunnel. Air flow was hitting the droplet from one side causing a significant disturbance to the shape of droplet. Therefore, there was some difficulty in recording the weight and temperature of the droplet. The transferred heat conduction to the droplet by the nozzle was another problem.

Cheong *et al.* [20] proposed a mathematical model to predict core temperature for drying a free suspended droplet against time. Reasonable agreement was obtained between the predicted and the experimental results at an air temperature of 20ºC. However, the predicted temperature was less accurate at higher temperatures (50ºC and 70ºC); the model was not applicable for cases at high air temperatures.

A mathematical model was modified in [21], taking into account the droplet shrinkage. The droplet was assumed first to undergo sensible heating with no mass change. The model showed that temperature distribution within the droplet cannot be ignored even for a small diameter droplet of 200 µm.

Wind tunnel definitely is considered one of the best tools to investigate and to study the role of boundary layer and the mechanisms of drying process. The most important variables in any drying process such as air flow, temperature and humidity arc usually easy to be controlled inside the wind tunnel. In the current study, through an experimental work and mathematical approach, we attempt to understand the role of the boundary layer on the interface behavior and the drying mechanisms for various materials of a flat plate surface and a single droplet shape.

2. BOUNDARY LAYER OVER A FLAT PLATE SURFACE

A boundary layer developed over a flat-plate surface plays a great role in the mechanisms of convective drying. Very little work has been done on the conjugated problem of heat and mass transfer during a flow over a drying bed. In this paper, through an experimental and mathematical approach, we attempt to understand the role of the thermal boundary layer on the interface behavior and the drying mechanisms for porous mediums. Beds of desert sand, beach sand and glass beads were subjected to forced convective drying in a scaled-up wind tunnel.

A laboratory-scale dryer designed for this work is shown in Figure 1. The apparatus consists of wind tunnel, molecular-sieve air dryer, 3 KW air heater and a fan. Through the wind tunnel a controlled flow of hot, dry air, with an average velocity of 1 m s^{-1}, was passed over a sample mounted flush with the tunnel floor. The last section of the wind tunnel (converging section) was designed to be opened easily for installation of the test bed.

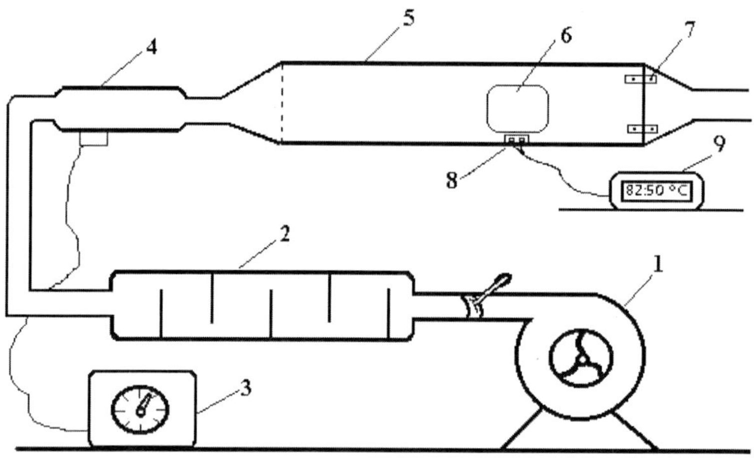

FIGURE 1. Experimental apparatus composed of: 1. fan; 2. molecular-sieve air dryer; 3. voltage regulator; 4. air heater; 5. wind tunnel; 6. observation port; 7. joint; 8. thermocouple socket; 9. thermocouple selector.

A glass tray (100 cm x 36 cm x 2.5 cm) attached to a flat metal plate was especially designed for this study. The top surface of the glass tray was at the same level of the metal plate that formed the floor of the wind tunnel. The sides and the bottom of the tray are insulated with neoprene rubber to minimize heat transfer *via* the glass wall. Figure 2 shows both the wind tunnel and the glass tray.

Ten thermocouples were inserted from one side of the tray to facilitate measurement of surface temperature distribution along the bed. At the same side of the tray, another ten thermocouples were inserted but at lower depth, 2.2 cm from the surface, to measure the bottom temperature distribution.

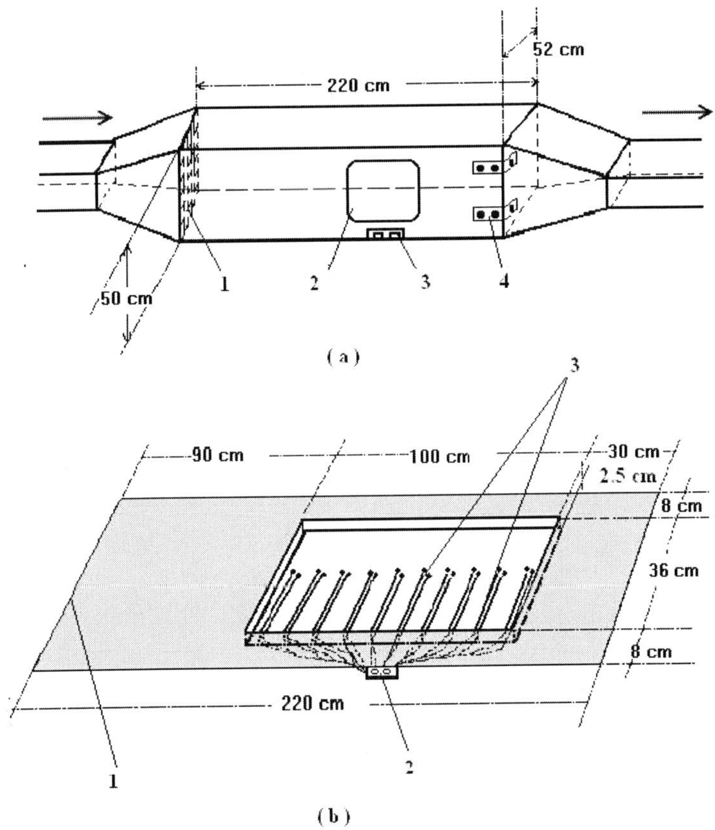

FIGURE 2. Details of (a) Wind tunnel: 1. smoothing grid; 2. observation port; 3.thermocouple socket; 4. joint, (b) Glass tray (clear) and flat plate (gray): 1. the leading edge; 2. thermocouple socket; 3. thermocouples.

The experiment was initiated by switching on the centrifugal fan and then the electric air heater. The voltage regulator was adjusted to provide the desired air temperature. The air temperature was monitored until it reached a steady state. This state was normally required 1 - 2 hours to be achieved.

When the apparatus achieved a constant air temperature, the drying process was commenced. The metal plate and the glass tray, containing the sample, were placed carefully inside the tunnel, allowing the hot air to pass over the surface of the bed. The initial readings of time and temperature were then registered immediately. During the experiment, temperature distributions were measured at intervals of approximately 20 minutes. The temperature at each distance was measured by using the thermocouple selector and registered with an accuracy ± 0.05ºC.

Two types of sand were subjected to the drying process. The first type was a fine sand of average grain size 220 μm diameter, with a moisture content of 0.17 kg kg^{-1}, taken from the desert of Kuwait, Burgan (N 28° 44 00 E 047° 42 00). The nature of the desert sand is usually fine and dry. The other type of sand was a beach sand of average grain size 300 μm diameter and 0.24 kg kg^{-1} moisture content, taken from the north coast of Kuwait, Bobiyan Island (N 29° 46 00 E 048° 22 00). The north coast is a place where the Shatt Al-Arab River (Iraq) falls into the Arabian Gulf. This makes the coast more muddy and less saline than the south coast. In addition, the beach sand contains a lot of small shells of various shapes. Glass beads of 400 μm diameter and 0.2 kg kg^{-1} initial moisture content were also selected, for test as a porous medium for comparison. The bed of glass beads is considered a typical sample for drying experiments, since the sizes of the beads are almost identical. The two types of sand and the glass beads beds are good examples for testing drying processes.

2.1. RESULTS

A bed of the desert sand was subjected to forced convective drying at 84 °C. The wet-bulb temperature of the air was 35.5 °C. The resulting surface temperature vs. time at different distances from the leading is shown in Figure 3, which demonstrates clearly the stages of the drying process; i.e. the pre-constant rate period, the constant rate period, and the falling rate period.

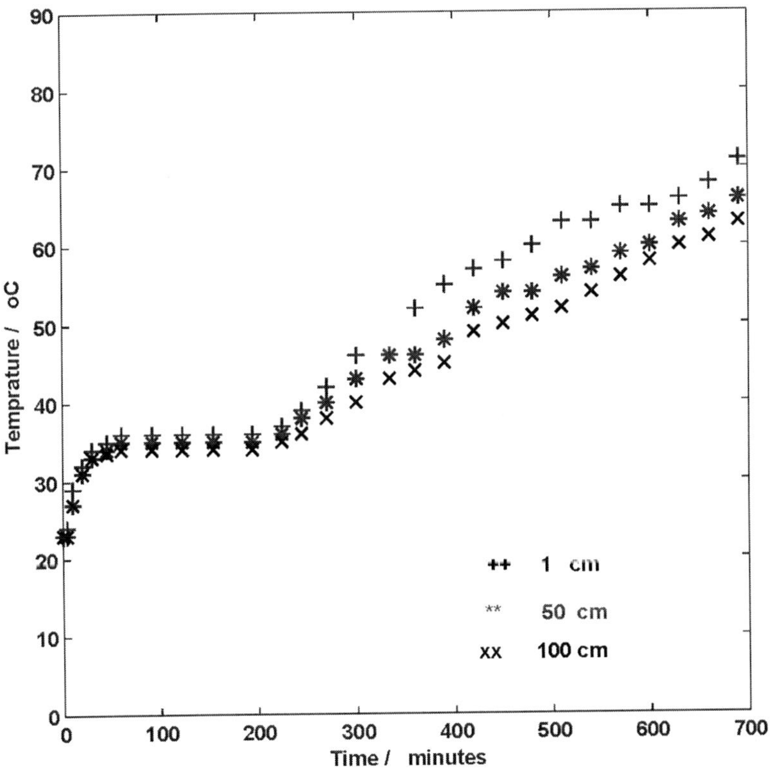

FIGURE 3. Temperature distribution profile of the surface for the desert sand bed.

The temperature of the bed rose from ambient temperature 23°C to a steady value at time = 70 minute. This initial period, termed the pre-constant rate period, is usually short. The surface temperatures remained constant for a period of 140 minutes, indicating the constant rate period. The surface temperature at the distance x=1cm, from the leading edge was 36 °C. This temperature was greater than that at x=50 cm and at x= 100 cm by 1 °C and 2 °C respectively. The surface temperatures were close to the wet-bulb temperature.

During the constant rate period, the surface of the solid is so wet that a continuous film of water exists on the drying surface. This water is entirely unbound water and exerts a vapour pressure equal to that of pure water at the same temperature. The rate of moisture movement within the solid is sufficient to maintain a saturated condition at the surface.

At a specific point, t = 240 minutes, the surface temperature at all positions rose gradually, indicating the end of the constant-rate period and the beginning of the falling-rate period. In the falling-rate period, there was insufficient water on the surface to maintain a continuous film. The entire surface was no longer wetted and dry patches began to form. The surface temperature continued to rise for a longer time until it approximated to the air temperature. A thin dried layer appeared on the entire surface.

The temperatures at the surface and the bottom at different distances from the leading edge are shown inFigure 4. The profiles show that when the surface became dry, the bottom remained wet at a constant temperature for a longer time than that for the surface. During the falling-rate periods a receding evaporation front divided the system into a hotter, dry zone near the surface and a wet zone towards the bottom of the sample [22]. The evaporation plane receded from the surface toward the bottom. The temperatures then rose quickly when the dry zone extended throughout the bed.

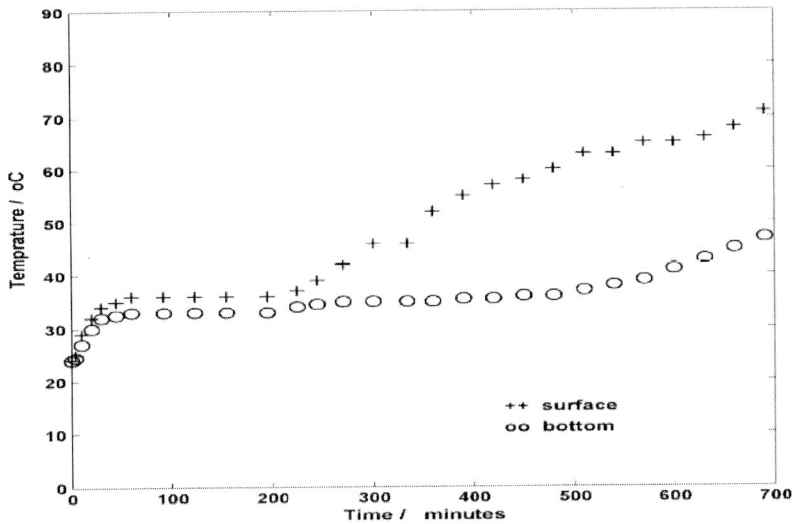
Temprature / oC
0
10
20
30
40
50
60
70
80
90
0
100
200
300
400
500
600
700
Time / minutes
++ surface
oo bottom

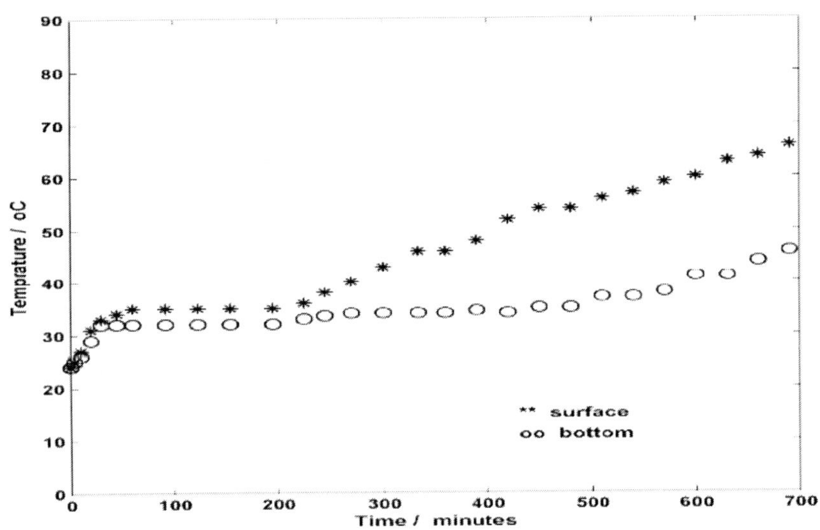
Temprature / oC
0
10
20
30
40
50
60
70
80
90
0
100
200
300
400
500
600
700
Time / minutes
** surface
oo bottom

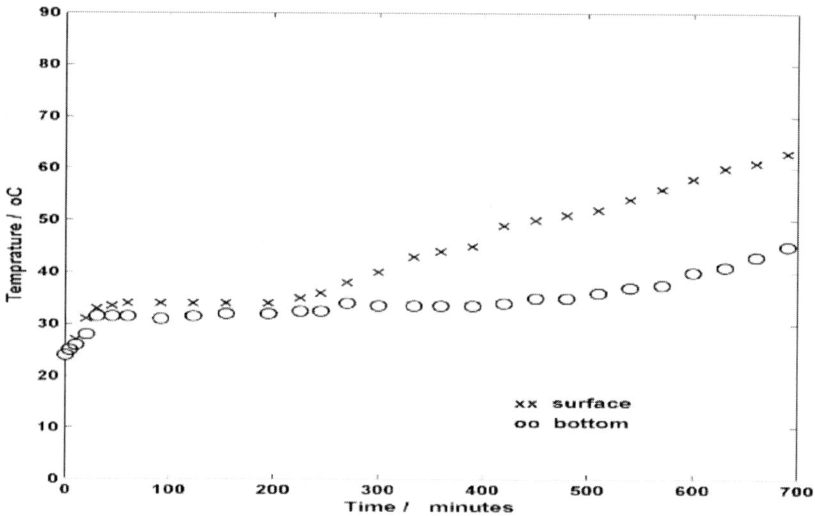

FIGURE 4. Temperatures of surface and bottom for the desert sand bed at distance of 1 cm, 50 cm and 100 cm from the leading edge, respectively, from the top.

A bed of the beach sand was dried at an air temperature of 83 °C. The wet bulb temperature was 35 °C. The temperature distribution profile at different distances from the leading edge is shown in Figure 5. The stages of the drying process can easily be recognized from the temperature profile. However, the surface temperature took longer to approach to the air temperature.

beach sand bed).

Figure 6 shows the surface and the bottom temperatures at distances from the leading edge of 1 cm, 50 cm 100 cm, respectively. It can be seen that the surface and bottom temperatures increased rapidly at some times and decreased others. Also, the temperature profile did not increase gradually like that of the desert sand. This can be attributed to the nature of the beach sand, which comprises different types of small shells of various shapes, and contain tiny hollows. The trapped water in these hollows forms small bubbles which can explode with increasing temperature. Therefore the temperature of the bed changes suddenly at such times. The temperature at the bottom of the bed indicated it remained wet during the falling-rate period.

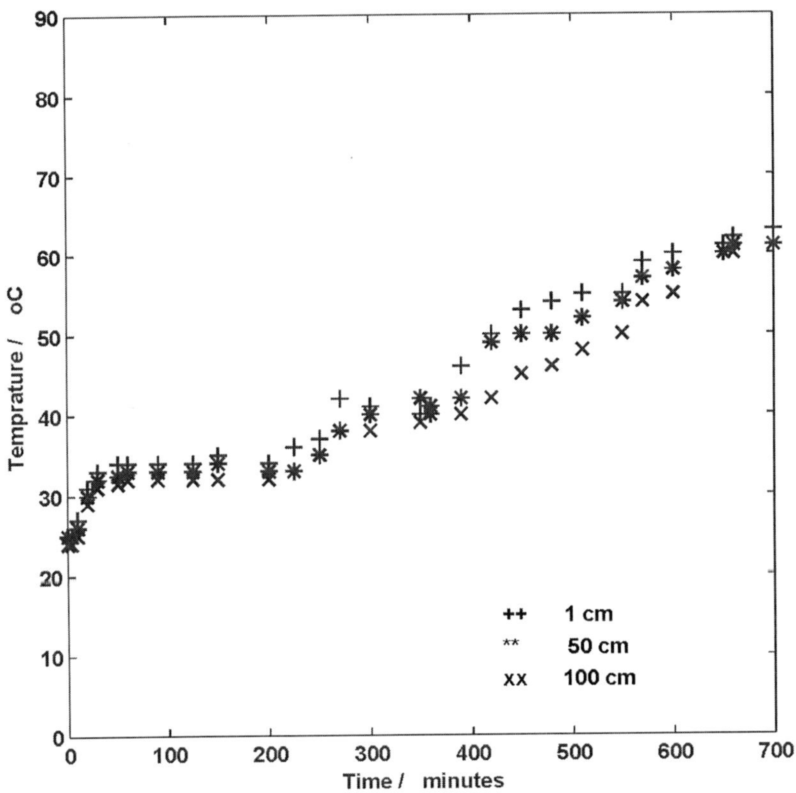

FIGURE 5. Temperature distribution profile of the surface for the

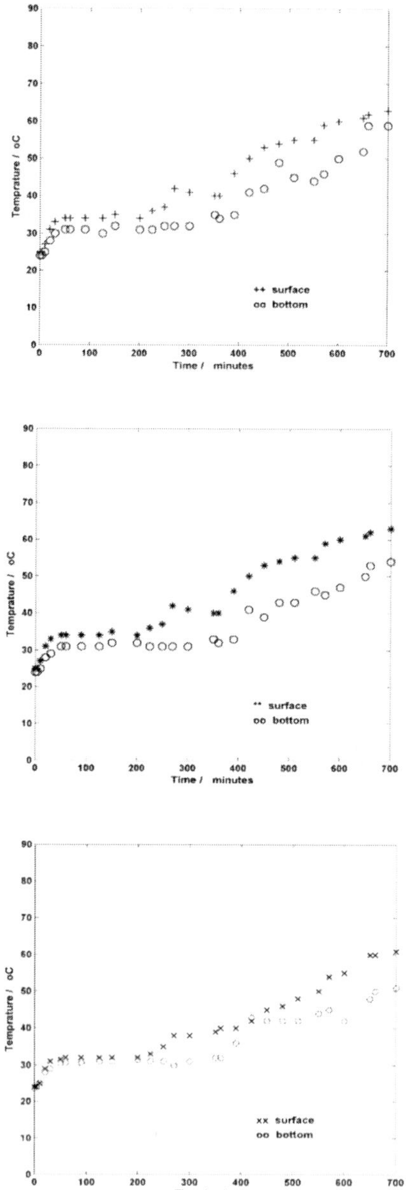

FIGURE 6. Temperatures of surface and bottom for the beach sand bed at distance of 1 cm, 50 cm and 100 cm from the leading edge, respectively, from the top.

A similar investigation was carried out for a bed of the glass beads at an air temperature of 84 °C. The wet bulb temperature was 37 °C. Figure 7 shows temperature versus time at different distances from the leading edge of the bed surface. The temperature distribution profile again illustrates clearly the stages of the drying process. The results for the surface and bottom temperatures for the glass beads were almost similar to those for the desert sand as shown in Figure 4.

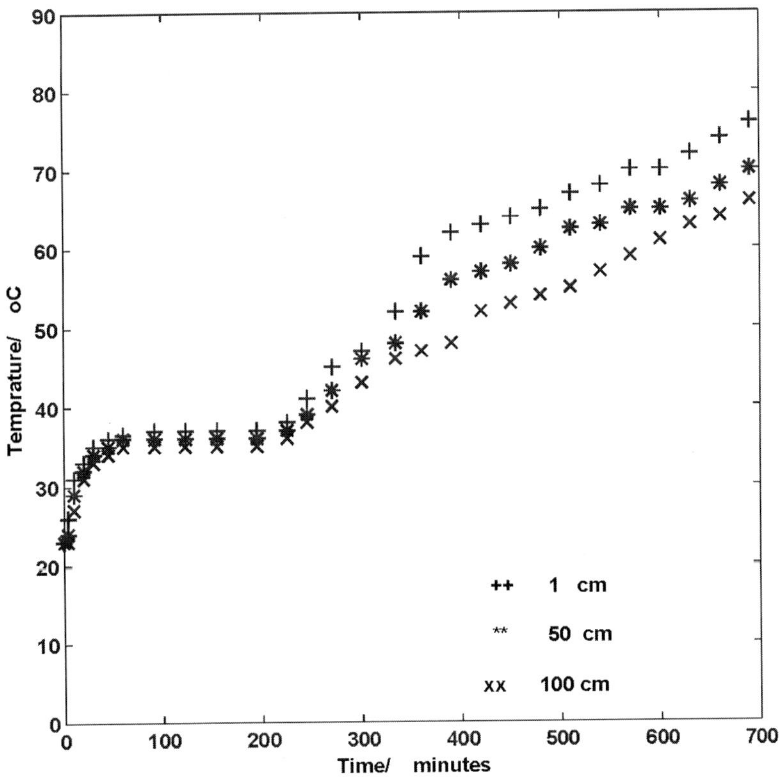

FIGURE 7. Temperature distribution profile of the surface for the glass-bead bed.

2.2. DISCUSSION

A thin film adjacent to the surface always exists when a forced flow passes over a flat plate and forms what is called a hydrodynamic boundary layer. The influence of the surface temperature reaches deeper into the fluid, thus causing the formation of a thermal boundary layer. It is well known, the thickness of the thermal boundary layer increases with increasing distance from the leading edge. This layer is affected by the geometry of the system, roughness of the surface and the fluid properties.

For the case in which the heated section is preceded by an unheated straight length, the local Nusselt number (Nu_x) is represented in [23-25] as:

$$Nu_x = \frac{hx}{k} = \frac{0.323(Re_x)^{0.5}(Pr)^{1/3}}{[1-(x_o/x)^{3/4}]^{1/3}} \tag{1}$$

where Rex is Reynolds number with respect to length and x is the length of the flat plate in (m).

Figure 8 shows a plot of variation in the local heat transfer coefficient versus the distance from the leading edge. The plot indicates that the values of the coefficient decreased significantly when x increased from the leading edge, and then it remains virtually constant for large x values. A plot of variation of the mass transfer coefficient could be expected to be almost similar to that in Figure 8because of the similarity in the transport coefficient equations. This result with the concept of the boundary layer thickness demonstrates that resistance to heat and mass transfer to, or from, the bed increases with increasing the distance from the leading edge.

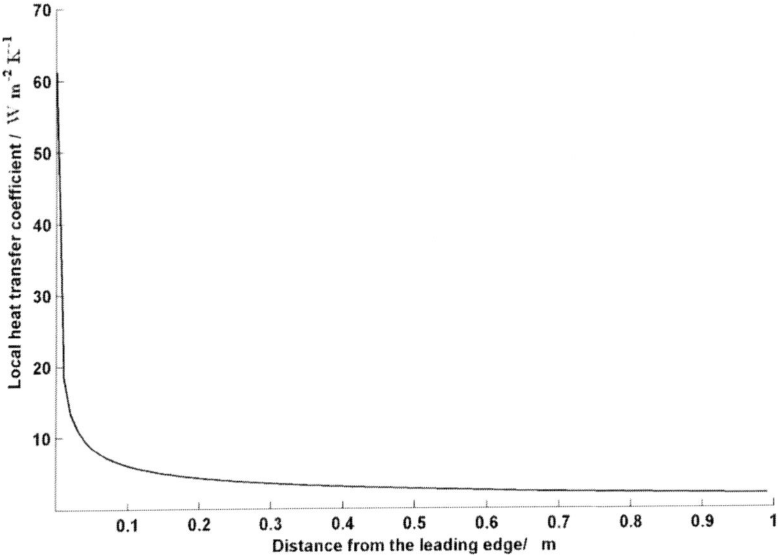

FIGURE 8. Local heat transfer coefficient vs. distance from the leading edge.

A model proposed in a previous paper [26] was modified to find a method for prediction of surface temperature distribution. The equation of energy can be represented as:

$$\varrho\, cp \frac{\partial T}{\partial t} = \nabla(k \nabla T) + \nabla(\frac{D_v M}{RT} \nabla P_v) \Delta hv \tag{2}$$

At time zero, the whole body has a uniform initial temperature of T_o, and the initial conditions are:

$$T_O = T\big|_{y=0} = T\big|_{y=h} \tag{3}$$

At the external surface, i.e. y=0, the boundary conditions can be written on the basis of Figure 9 as:

$$-k\frac{\partial T}{\partial y} = h(T_a - T_{sx}) + \dot{m}\,\Delta hv \tag{4}$$

where T_{sx} is the surface temperature at distance *x* from the leading edge.

The drying rate, m·m• varies with the time and can be defined as,

$$\dot{m} = \frac{M K'_G}{RT}(P_{sr} - P_v) \tag{5}$$

where K′GK′G is an overall mass transfer coefficient, defined in [27] as:

$$\frac{1}{K'_G} = \frac{1}{k'_c} + \frac{z''}{D_{eff}} \tag{6}$$

where k'ckc' is the local mass transfer coefficient (m s^{-1}), z"z" is the distance from the plate surface to the receding evaporation front in (m).

The boundary condition at *y* = h is:

$$-k\frac{\partial T}{\partial y} = 0 \tag{7}$$

Equation (2), with such boundary conditions, was solved by a finite difference method (a modified form the so-called explicit method). Therefore, the temperature distribution on the surface (i.e. *y* = 0) was calculated by using the model at different local points, *x*. Table 1 shows the physical properties of the sample materials.

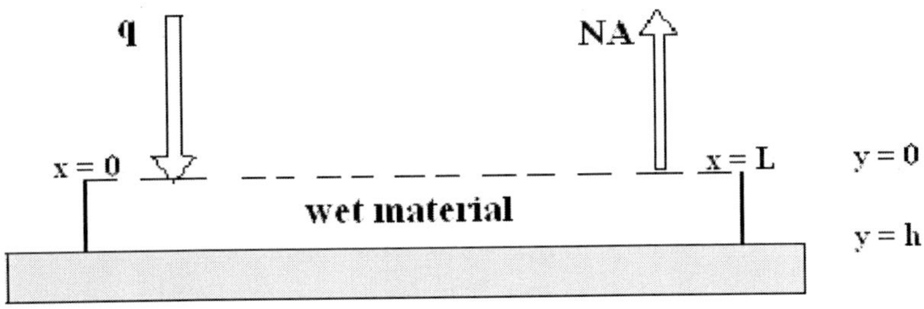

FIGURE 9. Heat and mass transfer process for the wet material. NA: mass transfer flux and q: heat transfer flux.)

Figure 10 shows the experimental and the predicted surface temperature distributions for the desert sand along the bed at various times. The predicted temperatures were in good agreement with the experimental results. From the graph, it can be seen that there is a significant difference in the surface temperature between 0.1 m and 1 m. At time=195 minutes; i.e. during the constant-rate period, the difference in temperature was 2 °C.

During the falling-rate period, the difference in the surface temperature between 0.1 m and 1 m increased. At 510 minutes, the difference is 11 °C. At the bottom of the sand bed, the difference in temperature along the bed also can be seen clearly (Figure 4).

For the entire time of the experiment, the surface and the bottom temperatures decreased gradually with increasing distance from the leading edge. This caused by the resistance to heat transfer process which increased with increasing thickness of thermal boundary layer. In contrast, near the leading edge, the resistance to heat transfer diminishes, since the thickness of the thermal boundary layer in the vicinity of the surface thins. Therefore, the rate of heat transfer to the body increased, thereby raising the temperature of surface. Afterwards, heat transfer by conduction across the solid particles raises the temperature of the bed, and the portion closest to the leading edge dries faster than that at a greater distance.

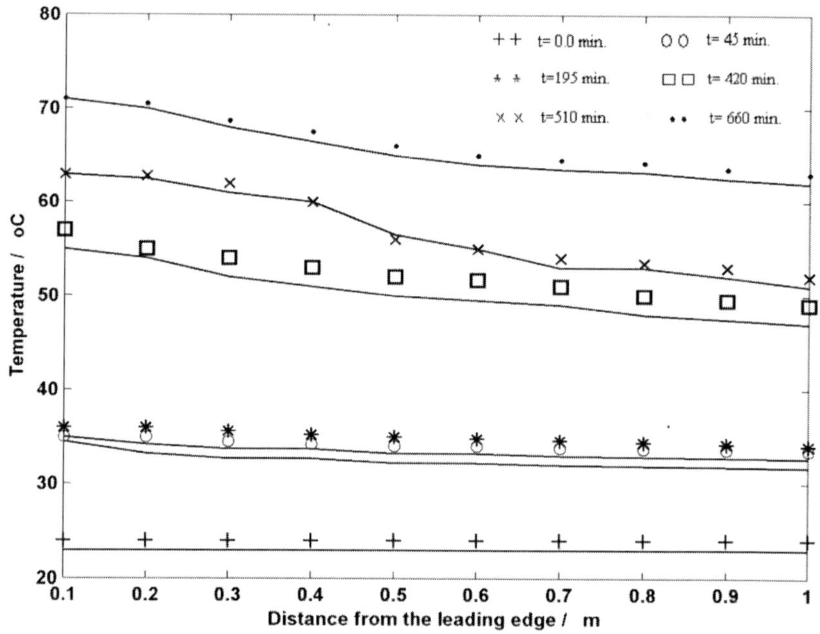

FIGURE 10. Experimental and predicted surface temperatures of the desert sand bed; experimental results (symbols); predicted results (solid lines).

Figure 11 shows both predicted and experimental results for the bed of beach sand. This Figure shows that the computed temperature distribution was in general agreement with experimental results. However, unsatisfactory results can be seen at times greater than 400 minutes, i.e. during the falling-rate period. This is due to the nature of the beach sand as discussed before. The surface temperature of the beach sand sample decreased gradually with increasing distance from the leading edge. A significant difference in temperature between 0.1 m and 1 m can be seen clearly for both constant and falling-rate periods. During the constant-rate period, the difference in temperature was 2 °C, whereas in the falling-rate period the difference reached to 8 °C.

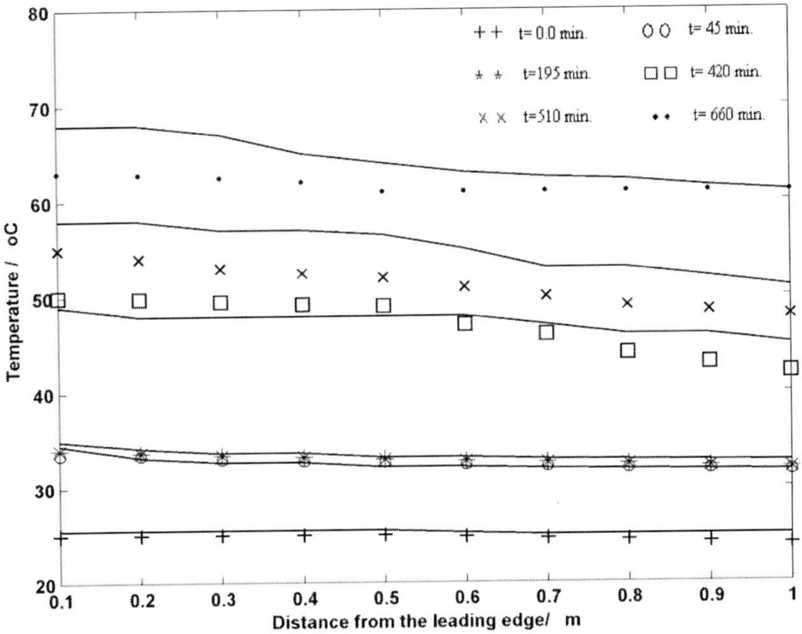

FIGURE 11. Experimental and predicted surface temperatures of the beach sand bed; experimental results (symbols); predicted results (solid lines).

TABLE 1. Physical properties for desert sand and beach sand.

	ρ(Kg/m^3)	cp_d (kJ/kg K)	k (W/m K)	k_d(W/m K)	D_{eff} (m^2/s)	D_v (m^2/s)
Desert sand	1760	1.1	0.760	0.246	0.81x10^{-5}	1.75x10^{-5}
Beach sand	1980	1.2	1.12	0.188	1.15x10^{-5}	2.05x10^{-5}

We have found that the temperature distribution profiles determined for the flat beds of desert sand, beach sand and glass beads identified clearly the stages of drying. The temperature profiles in general were almost similar. However, the beach sand profile showed irregularity in temperature, due to the nature of the beach sand which contains a lot of shells with various shapes.

The temperature profiles also showed that when the whole surface of the bed became dry, the whole bottom of the bed remained wet. During the falling-rate periods, a receding evaporation front divided the system into a hotter, dry zone near the surface and a wet zone towards the bottom of the sample.

The predicted transport coefficients have very large values close to the leading edge, where the thickness of the boundary layer approaches zero. In contrast, the values of the coefficients decrease progressively with increased distance from the leading edge, where the boundary layer thickens. Hence the resistance to heat and mass transfer to, or from, the surface also increases. These variations in thickness and resistance have a significant effect on the temperature distribution along the bed and the drying rate.

A mathematical model has been modified to predict temperature distributions along the bed at various times. The model was compared with the experimental results for various beds and good agreement was obtained. We found that surface and bottom temperatures decreased gradually with increasing distance from the leading edge, and the difference in temperature became clearer during the falling-rate period. The difference in the surface temperature was 11 °C for the case of desert sand, and was 8 °C for the case of beach sand. We concluded that the portion close to the leading edge dried faster than that at larger distance, since the resistance to heat and mass transfer diminishes at that position.

3. SINGLE DROPLET DRYING

The study of mechanisms that describe single droplet drying is a challenging issue since it involves many disciplines: heat and mass transfer, fluid mechanics and chemical kinetics. The main objective of this section is to attempt to further understand the mechanisms involved in the drying of a single droplet, and more specifically to formulate a mathematical model. The model should predict temperature profiles for both the inner core and the outer surface of the droplet under simulated conditions that might be encountered in spray drying equipment. Experimental work will also involve the measuring the moisture contents against time to provide more information for the droplet drying process.

The experimental apparatus was comprised of a horizontal wind tunnel 2.2 m long. The wind tunnel supplied a forced drying air into the working section where the droplet was suspended from a glass nozzle. The apparatus and the flow system are shown in Figure 12. A gate valve at the inlet to the wind tunnel controlled airflow rate. Air was heated to the desired temperature, using a 3 KW electric heating element controlled by a rotary voltage regulator.

Through the wind tunnel, a controlled flow of hot, dry air, with an average velocity of 1 m/s, was passed across the droplet suspended from the glass nozzle. A forced dry air was obtained by using a centrifugal fan and a molecular sieve air dryer containing silica gel and calcium silicate.

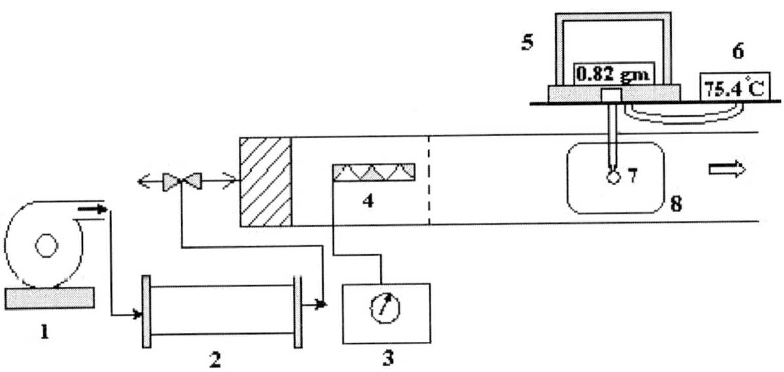

FIGURE 12. Experimental apparatus: 1-fan 2-molecular sieve 3-voltage regulator 4-air heater 5-Digital balance 6-Temperature recorder 7-Glass nozzle 8-Observation port.

The experiment was initiated by switching on the centrifugal fan and then the electric air heater. The voltage regulator was adjusted to provide the desired air temperature. A thermocouple was fixed in the wind tunnel near to the glass nozzle to measure the air temperature with an accuracy of ± 1° C. The air temperature was monitored until it reached a steady state. This state requires between 1 to 2 hours to be achieved. When the apparatus achieved a constant air temperature, the drying process was initiated.

As a part of this study, a droplet suspension device was specially designed to measure the weight and temperature of the droplet. The droplet suspension device is illustrated in Figure 13. It consisted of a glass nozzle with the dimensions of 180 mm in length and 9 mm outside diameter. The upper section of the glass nozzle was fixed by a small electric rotator device able to provide a rotational speed range of 1- 30 rpm. In order to reduce the contact area between the free end of the glass nozzle and the droplet surface, the lower section of the nozzle was shaped as a small cone with a free end diameter of 4 mm. The droplet receives a relatively small amount of heat transferred by conduction through the glass nozzle, which has a thermal conductivity of 0.480 W/m K. The heat transferred by conduction was taken into account in the proposed model represented by Equations 4&5. The glass nozzle was rotated at constant low speed of 5 rpm. This rotation speed had no effect on the shape or stability of the droplet. The upper section of the suspension device was fixed by a metallic clip installed beneath the analytical balance. The lower section of the suspension device, i.e. the glass nozzle, was inserted through a hole in the wind tunnel.

Two thermocouple sensor types, NiCr-NiAl, were used to measure the core and surface temperatures of the droplet. One of the thermocouples was placed inside the glass nozzle and extended to the center of the droplet. The other thermocouple was fixed outside and along the glass nozzle. The outer thermocouple rotates simultaneously with the rotation of the glass nozzle. The end tip of the thermocouple was positioned in a manner to touch the outer surface of the droplet. The rotation process that made both the nozzle and the droplet rotate together assisted in avoiding any separation between them that might have been caused by the force of air drying. The core and the surface temperatures of the droplet

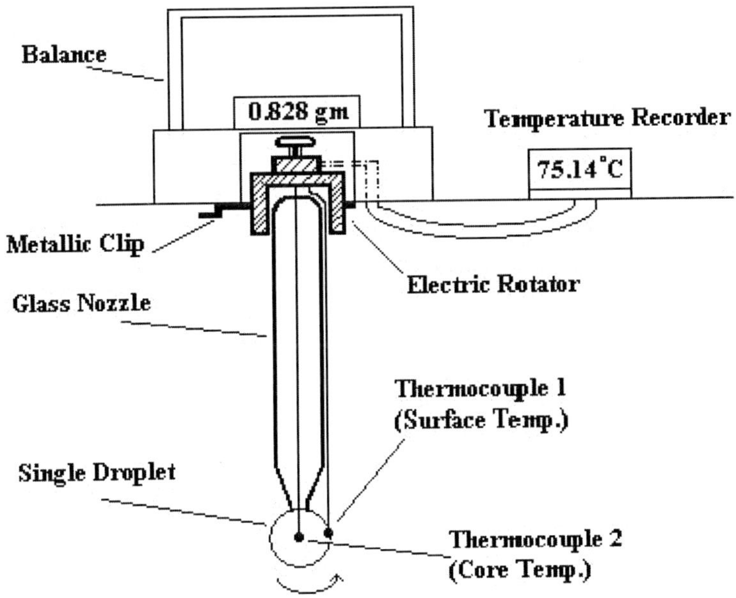

FIGURE 13. Droplet suspension device; glass nozzle and the connected thermocouples.

were easily recorded by a temperature recorder at 50 sec. intervals. The drying process of the different material droplets was investigated under air temperature of 75° C and 140° C.

The procedure for weighing the droplet was carried out quickly and intermittently by causing the suspension device to be freely-suspended. The gate valve was closed to cut off the airflow to the working section and diverted to an outlet 20 mm valve in order to prevent any vibration of the glass nozzle during the weighing process. The nozzle rotation was simultaneously stopped. A metallic clip was opened manually to allow the suspension device to be freely-suspended from a hook connected beneath the balance. This arrangement made the weight measurement readings more accurate. The weighing procedure was repeated during the droplet drying experiment at 100 secintervals. Thus, the weight loss of droplet was recorded and moisture content was determined versus time. The required time for each weighing procedure step was about 10 sec. This time was not included in the recorded drying time and had no noticeable effect on the results.

Three types of liquids were selected for the drying process experiments. The first type was sodium sulphate decahydrate solution (60 wt % solid). The second type was a concentrated fruit juice (60 wt % fruit juice powder of apple, peach and blueberry, MTC product). The third type was an organic paste (20% sodium chloride, 25% dispersal pigment; Goteks product) used for adhesive and coating applications. Table 2 shows the physical properties of the sample materials. Droplets ranging from 9 mm to 14 mm diameter were subjected to the drying process. However, the actual size of the droplets in typical spray drying applications is much smaller. The droplets with small sizes require developing a more accurate technique to measure both core and surface temperatures. Therefore, the current research assumes that the mechanisms of the drying large and small droplets are similar.

TABLE 2. Physical properties of sodium sulphate decahydrate solution, fruit juice and organic paste.

	ρ_d (Kg/m^3)	cp_d (kJ/kgK)	k (W/mK)	k_d (W/mK)	D_{eff} (m^2/s)	D_v (m^2/s)
Sodium sulphate decahydrate solution	3110	1.1	0.180	0.246	1.14×10^{-5}	3.45×10^{-5}
Fruit juice	1650	1.7	0.126	0.188	1.10×10^{-5}	3.40×10^{-5}
Organic paste	3400	2.4	0.251	0.422	1.52×10^{-5}	3.50×10^{-5}

3.1. MATHEMATICAL MODEL

In the drying process, the droplet is first heated by the hot air flow with significant evaporation from the surface. The temperature of the surface increases and approaches the wet-bulb temperature, indicating the constant drying rate period. In order to propose a mathematical model, the droplet was assumed to have a fixed size with no change during this period. Also, the droplet was assumed to have a uniform initial temperature and moisture content. Temperature distribution within the droplet can be represented as

$$\frac{\partial T}{\partial t} = \alpha\left(\frac{\partial^2 T}{\partial r^2} + \frac{2}{r}\frac{\partial T}{\partial r}\right) \tag{8}$$

Equation 8 was solved with the following boundary conditions using explicit finite differences,

$$-k\frac{\partial T}{\partial r} = 0 \qquad at\ r = 0 \tag{9}$$

and

$$-k\frac{\partial T}{\partial r} = h(T_a - T_{sr}) + q_{nz} \qquad at\ r = R \tag{10}$$

where, q_{nz} is the transferred heat conduction to the droplet through the glass nozzle. It can be calculated as:

$$q_{nz} - \left(\frac{4k_{nz}h_{nz}}{d_{nz}}\right)^{0.5}(T_a - T_{sr}) \tag{11}$$

The heat transfer coefficient, h_{nz}, can be correlated from Thomas (1999) as:

$$h_{nz} = 0.26 + Re^{0.6}Pr^{0.33}\left(k_{air} / d_{nz}\right) \tag{12}$$

During the falling rate period, the formation of a receding evaporation front divides the droplet into two regions, a dry crust at the outer surface and a wet region inside the core. Therefore, heat transfer equations are formulated for each region. The physical model and the coordinate system for analysis are shown in Figure 14.

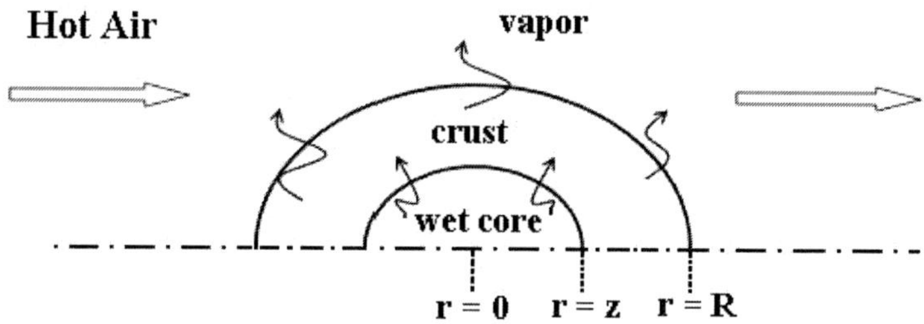

FIGURE 14. Physical model and the coordinate system of the droplet cross-section.

Energy balance for the wet core, 0 < r < z, can be represented as follows:

$$\frac{\partial T_w}{\partial t} = \alpha_w \left(\frac{\partial^2 T_w}{\partial r^2} + \frac{2}{r} \frac{\partial T_w}{\partial r} \right) \tag{13}$$

Heat is transferred through the crust into the wet core where evaporation occurs at the interface between the core and the crust. Vapor then diffuses through pores of the crust into the drying medium. Thus, moisture is transferred mainly by vapor flow. Consequently, vapor diffusion must be taken into account in formulating the equations for the dry (crust) region. The energy balance for the crust region, z < r < R, can be represented as follows:

$$\frac{\partial T_d}{\partial t} = \alpha_d \left(\frac{\partial^2 T_d}{\partial r^2} + \frac{2}{r} \frac{\partial T_d}{\partial r} \right) + \frac{\partial}{\partial r} \left(\frac{M}{RT} \frac{\partial P_v}{\partial r} \right) \frac{\Delta h_v}{\rho_d c p_d} D_v \tag{14}$$

3.2. BOUNDARY CONDITIONS

At the center of the sphere, r = 0

$$-k_w \frac{\partial T_w}{\partial r} = 0 \quad (15)$$

At the surface, r = R:

$$-k_d \frac{\partial T_d}{\partial r} = h(T_a - T_{sr}) + q_{nz} \quad (16)$$

At the receding evaporation front (r = z), the moving boundary conditions are

$$-k_d \frac{\partial T_d}{\partial r} + k_w \frac{\partial T_w}{\partial r} = \dot{m}\,\Delta hv \quad (17)$$

The drying rate, m·m•, was defined in Eqs. (5 & 6). However, in this case, z″z" , in Eq. (6) represents the distance from the droplet surface (r = R) to the receding evaporation front (r = z).

Heat and mass transfer coefficients, h and k′ckc' , can be determined by the correlations found in [28,29] as follows:

$$Nu = 2 + \Phi Re^{0.5} Pr^{0.33} \quad (18)$$

and

$$Sh = 2 + \beta Re^{0.5} Sc^{0.33} \quad (19)$$

where Φ and β are constants ranging from 0.6-0.7 for Re (500 - 17000).

Thermal conductivity of the wet region can be evaluated according to [30] as:

$$k_w = k_d + k_v \overline{X} \quad (20)$$

where $\overline{X}$ is an average moisture content and k_v is

$$k_v = \frac{D_v M}{RT} \cdot \frac{dp}{dT} \Delta h_v \quad (21)$$

The non-linear Eqs. (13&14) with the boundary conditions as in Eqs. (15-17) were solved by a finite difference method. The proposed equations were solved in a program using Turbo-Pascal V.6, and the computational results compared with the experimental results.

3.3. RESULTS

A single droplet of sodium sulphate decahydrate solution was suspended from the free end of the glass nozzle at an air temperature of 75° C and an air velocity of 1 m/s. The wet-bulb temperature was 46° C. The core and surface temperatures of the droplet versus time are plotted in Figure 15. Initially, the temperatures increased rapidly because of the large difference in temperature between the drying medium and the droplet. Basically, the plot exhibited two drying periods. A short period, from time =100 - 200 s, where the surface temperature approached the wet-bulb temperature, is described as the constant rate period. The second period, or falling rate period, extended from t =200 s to the end of the experiment.

Figure 15 plots the predicted values of both the core and surface temperature evaluated by the developed model. Comparison of the experimental and theoretical results showed good agreement. However, actual surface temperatures were slightly higher than those predicted by the model. This probably was due to the position of the thermocouple and its reading during the experiment, as will be discussed later in more detail.

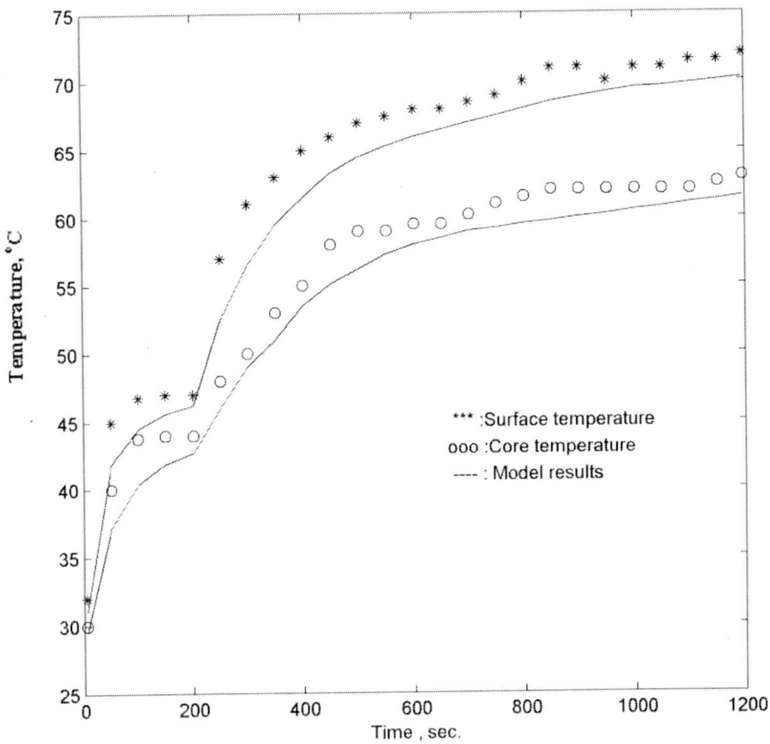

FIGURE 15. Temperature distribution profile for sodium sulphate decahydrate at air temperature of 75° C.

The previous experiment was repeated at an air temperature of 140° C in the same drying medium.Figure 16 shows experimental and theoretical results of a droplet of the same material. The experimental results showed that the constant rate period extended from time= 85 - 165 s, shorter than that observed at 75° C. The results showed both surface and core temperatures of 130° C at 900 s, at which point the droplet had dried completely. Figure 16 showed agreement between the experimental and predicted results.

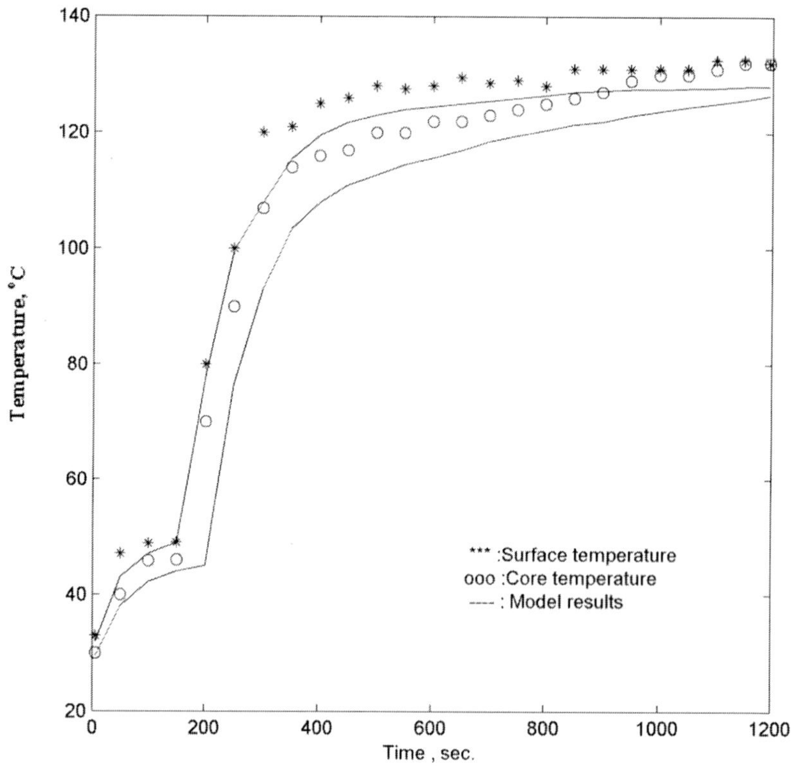

FIGURE 16. Temperature distribution profile for sodium sulphate decahydrate at air temperature of 140° C.

A single droplet of a fruit juice was dried at an air temperature of 75° C and an air velocity of 1 m/s. Wet-bulb temperature was 44.5° C. The experimental results in Figure 17 show a constant rate period from time =100 - 200 s. The falling rate period can be observed after the constant period, when the temperature increased quickly.

During the falling rate period some experimental readings represented approximate values of the actual droplet surface temperatures. The approximate values can be attributed to the fact that the droplet diameter constantly decreased with time and that caused the distance between the tip of thermocouple and the surface of droplet to grow. In other words, during some time in the experiment, a part of the end tip of the thermocouple was touching the surface of the droplet, and the remaining area of the tip was exposed to air flow. Such a behavior caused the tip of thermocouple to give an average reading for both the surface and the air flow temperatures. Good agreement was obtained between the theoretical and experimental results of the temperature profile for the fruit juice.

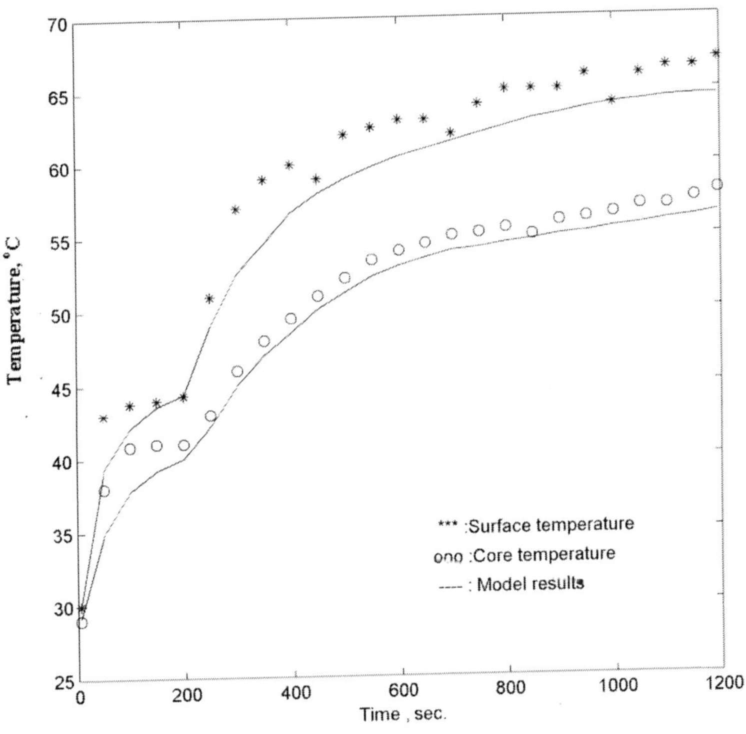

FIGURE 17. Temperature distribution profile for fruit juice at air temperature of 75° C.

The fruit juice droplet was dried again at an air temperature of 140° C in the same drying medium. The results (Figure 18) showed that the constant rate period was a little shorter than that at 75° C. The results showed also that both surface and core temperatures were similar and close to the air temperature at time = 1050 s. The mathematical model showed also good agreement with the experimental readings.

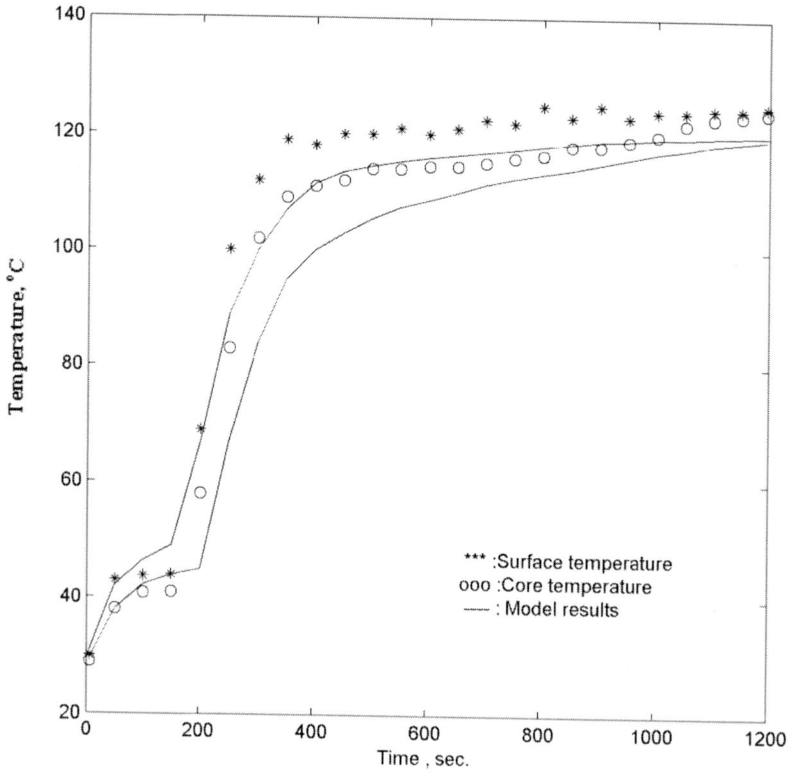

FIGURE 18. Temperature distribution profile for fruit juice at air temperature of 140° C.

A single droplet of organic paste was dried under similar conditions to that of the fruit juice. A plot of the core and surface temperatures of the droplet versus time is shown in Figure 19. A relatively longer period of constant drying rate, compared to those shown for sodium sulphate and fruit juice, was observed. Figure 19 also showed that surface temperature increased rapidly after time = 300 s, indicating the beginning of the falling rate period. Less agreement was obtained between the model's predicted results and the experimental results. Organic paste, which has higher thermal conductivity, forms a thicker solid crust at the outer surface compared to the other material evaluated. Therefore, a higher resistance to heat and vapor through the crust would be expected. To improve the predictions, introduction of a correction factor is probably required in the model for those materials which have a nature similar to that of the organic paste. Figure 20 shows the experimental and theoretical results of air temperature of 140° C. Again, forming a solid crust at the outer surface led to less agreement between theoretical and experimental results. The predicted temperature distributions were higher than the actual temperatures.

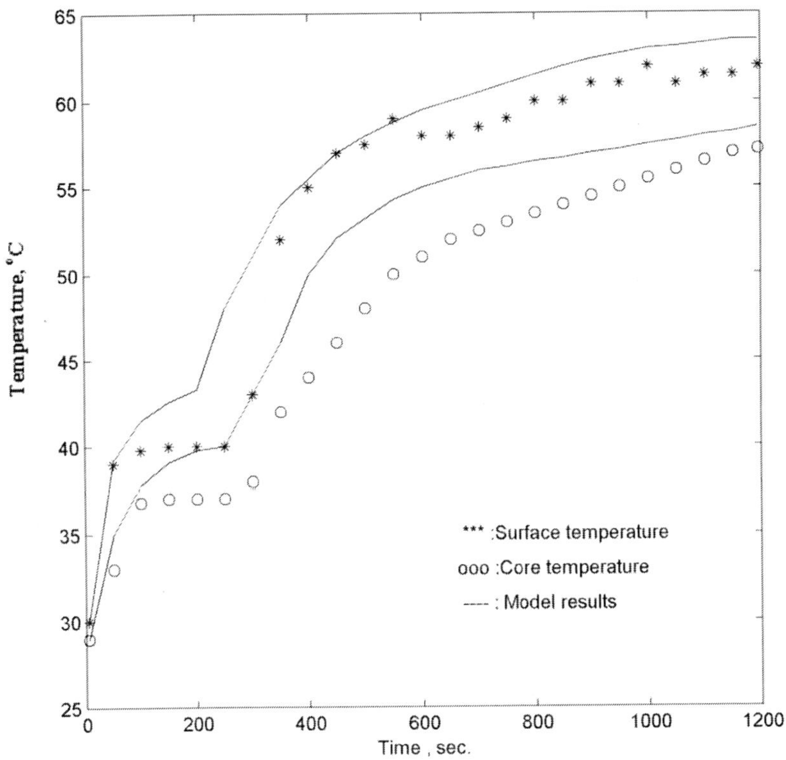

FIGURE 19. Temperature distribution profile for organic paste at air temperature of 75° C.

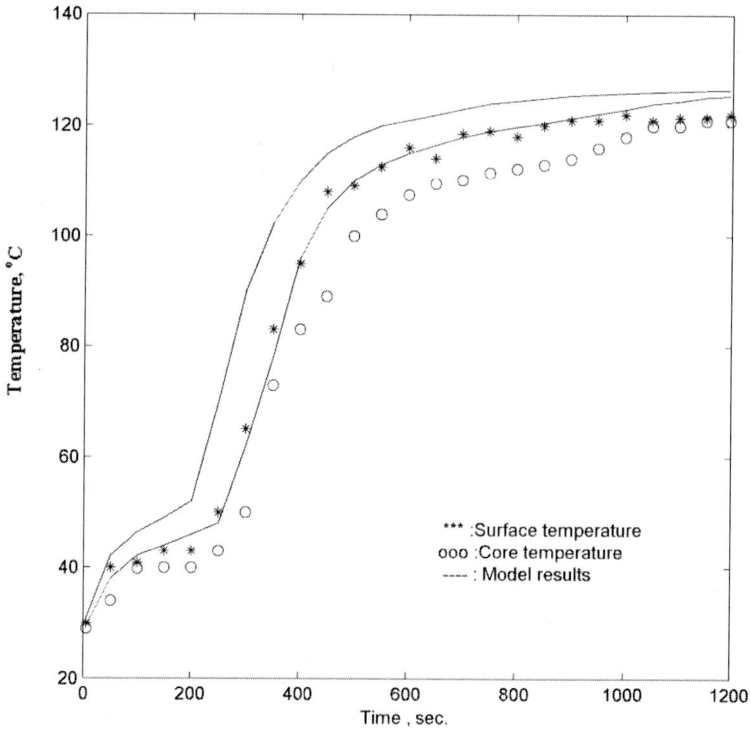

FIGURE 20. Temperature distribution profile for organic paste at air temperature of 140° C.

The moisture content of the droplet was determined by measuring the weight loss against time. There was no experimental technique to measure core and surface moisture content separately; therefore, the measured moisture content of the droplet represented the average value. The experimental results of moisture content distribution for the three samples at an air temperature of 75° C, are shown in Figure 21. The moisture content profiles clearly show the two stages of drying, the constant rate period and the falling rate period. The profiles show that the sodium sulphate decarbohydrate solution had consistently lower moisture content, as it dried faster than the other materials. The profiles also showed that forming a solid crust in the falling rate period lowered the moisture content values for all the three samples. In the case of organic paste, the change in the moisture content was more significant.

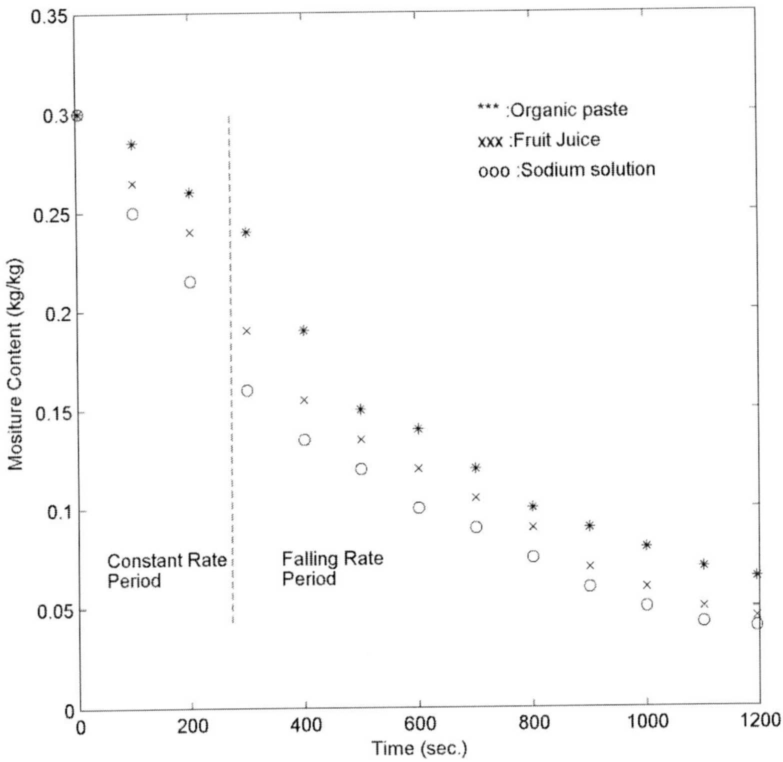

FIGURE 21. Moisture content profile for all samplcs at air temperature of 75° C.

Obviously, crust formation, thickness and porosity have a significant effect on the moisture content and on the drying rate of the droplet. This result was also obtained by Hayder & Mumford [31] in the drying of custard and starch droplets. They observed that crust formation was more rapid on the custard droplet, because the smaller starch granules absorbed less and left more free water in the droplet. Therefore, the crust growth and the drying rate of the starch droplets took a longer time.

Moisture distribution curves for the three samples at 140° C air temperature are plotted in Figure 22. Similar results were obtained to those observed at an air temperature of 75° C. Also, the profiles showed that the moisture content dropped to lower values compared to those at 75° C. In other words, the droplets dried faster at the higher temperature.

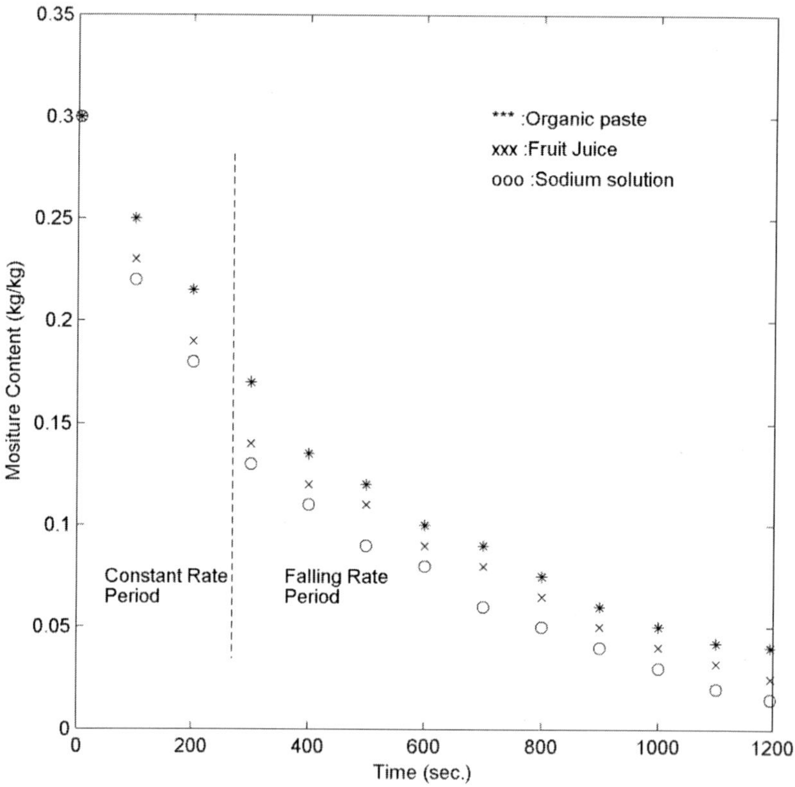

FIGURE 22. Moisture content profile for all samples at air temperature of 140° C.

3.4. DISCUSSION

As previously observed, moisture and temperature distribution profiles of various materials exhibited constant and falling drying rate periods. In the constant rate period, the temperature of the droplet surface was almost equal to the wet-bulb temperature. During this period, evaporation takes place from the free liquid surface of the droplet. The constant rate period was relatively short in sodium sulphate and fruit juice samples. However, that period was longer in the case of organic paste.

The falling rate period is characterized by formation of a partial crust on the outer surface of the droplet. This crust recedes towards the core and the surface temperature starts to increase. Vapor diffusion becomes the predominant transport process at this stage. Crust structure, thickness and porosity have a significant effect on the rate of drying. The crust thickness increases with time, hence the resistance to heat and moisture diffusion through the crust increases. Therefore, moisture content and drying rate decrease.

Some experimental readings represented approximate values of the actual droplet surface temperatures. This was attributed to the end tip of thermocouple that was giving average readings for both the surface and the air flow temperatures at the same time. It was also assumed that the mechanism of the droplet drying is similar for both small and large diameters. Therefore, it is recommended that a more accurate technique for measuring droplets with small sizes and for the taking of surface temperatures be developed.

The new model predicted temperature distribution profiles for single droplets of various materials. The predicted results showed a good agreement with the experimental data for air temperatures at 75° C and 140° C. However, the model was less accurate in the case of organic paste due to the higher thermal conductivity of the formed crust. A correction factor should be developed and taken into account for such materials. The model provides a relatively fast and efficient way to simulate drying behavior over a range of drying conditions. The model also represents a useful tool in the design and optimization of spray drying processes.

Moisture content profiles clearly showed the two stages of the drying process. In addition, the moisture profiles supported the conclusions that the crust forming in the falling rate period decreased both the moisture content and the drying rate.

Through the results of the experimental work and the theoretical model for a droplet drying, a significant conclusion was obtained. It has always been wrongly assumed in the literature that there is no temperature distribution within the droplet. This concept has been corrected by the current research. All experiments for the three materials used showed a clear difference between the core and the surface temperatures of the droplet during the drying process.

3.5. GENERAL CONCLUSION

The characteristics of the boundary layer have a great effect on the local heat and mass transfer coefficients and temperature distributions throughout a flat bed surface. Droplet drying is an important subject in drying science, since it provides more details about the drying mechanisms in order to optimize spray dryer equipment. However, this part of the drying field is rarely considered in the literature. Further research in this area seems essential to obtain better understanding of drying theory.

Wind tunnel definitely is considered one of the best tools to investigate and to study the mechanisms of drying process. The most important variables in any drying process such as air flow, temperature and humidity are usually easy to be controlled inside the wind tunnel. Through a mathematical approach and an experimental work using a wind tunnel, we highlighted on the role of the boundary layer on the interface behavior and the drying mechanisms for various materials of a flat plate surface and a single droplet shape.

NOMENCLATURE

α	Thermal diffusivity	m^2/hr
Cp	Heat capacity	kJ/kg K
D	Diameter	M
D_{eff}	Effective diffusivity	m^2/s
D_v	Diffusivity of vaporization	m^2/s
H	Heat transfer coefficient	W/m^2 K
Δh_v	Latent heat of vaporization	kJ/kg
K	Thermal conductivity	W/m K
k'_c	Mass transfer coefficient	m/s
K'_G	Overall mass transfer coefficient	m/s
M	Molecular weight of water	kg/kg_{mol}
Nu	Nussult number	-
P	Partial pressure	kPa
Pr	Prandtel number-	

Q	Conduction heat transfer	W/m^2
R	Universal constant	m^3kPa/kg_{mol} K
Re	Reynolds number	-
t	Time	s
T	Temperature	K
X	Moisture content	kg/kg
T_a	Air Temperature	K
ρ	Density	kg/m^3
Sc	Schmidt number	

SUBSCRIPTS

a	Dry crust
nz	Glass nozzle
sr	Surface
v	Vapor
w	Wet core

REFERENCES

1. Sparrow E, Lin S. Boundary Layer With Prescribed Heat Flux Application to Simultaneous Convective and Radiation. International Journal of Heat and Mass Transfer 1965; 8, 437- 448.

2. Luikov A. Conjugate Convective Heat Transfer Problems. International Journal of Heat and Mass Transfer 1974; 17, 257-265.

3. Chyou T. The effect of A Short Unheated Length and A Concentrated Heat Source on The Heat Transfer Through A Turbulent Boundary Layer. International Journal of Heat and Mass Transfer 1991; 34, 1917-1928.

4. Harris S, Ingham D, Pop I. Transient Boundary Layer Heat Transfer From A Flat Plate Subjected To A Sudden Change in Heat Flux. European Journal of Mechanics B- Fluid 2001; 20, 187-204.

5. Deswita L, Nazar R, Ahmad R, Ishak A, Pop I. Similarity Solutions of Free Convection Boundary Layer Flow on a Horizontal Plate with Variable Wall Temperature. European Journal of Scientific Research 2009; 27(2) 188-198.

6. Defraeyea T, Houvenaghelc G, Carmelieta J, Derome D. Numerical Analysis of Convective Drying of Gypsum Boards. International Journal of Heat and Mass Transfer 2012; 55, 4487-4928.

7. Mori S, Nakagwa H, Tanimoto A, Sakakibara M. Heat and Mass Transfer With A Boundary Layer Flow Past A Flat Plate of Finite Thickness. International Journal of Heat and Mass Transfer 1991; 34, 2899-2909.

8. Masmoudi W, Prat M. Heat and Mass Transfer Between A Porous Medium and A Parallel External Flow; Application to Drying of Capillary Porous Material. International Journal of Heat and Mass Transfer 1991; 34, 1975-1989.

9. Jomaa W, Bruneau D, Nadeau J. Simulation of The High Temperature Drying of A Past Product: on The Influence of The Local Air Flow and The Thermal Radiation. Drying Technology 2004; 22, 1709-1729.

10. Ranz W, Marshall W. Evaporation From Drops. Chemical Engineering Progress 1952; 48, 141-173.

11. Trommelen A, Crosby E. Evaporation and Drying of Drops in Superheated Vapors. AIChE Journal 1970; 16, 857-872.

12. Sirignano, W.A. Fluid Dynamics and Transport of Droplets and Sprays: Cambridge University Press, NY, USA 1999.

13. Masters K. Spray Drying in Practice. Spray Dry Consult International: Denmark: John Wiley & Sons Inc., NY, USA 2002.

14. Sloth J, Kiil S, Jensen A, Andersen S, Jørgensen K, Schiffter H, Lee G. Model Based Analysis of The Drying of A Single Solution Droplet in An ultrasonic Levitator. Chemical Engineering Science 2006; 61, 2701-2709.

15. Miura K, Miura T, Ohtani S. Heat and Mass Transfer to and From Droplets. American Journal of Chemical Engineers, Symposium series 1977; 73 (163).

16. Ali H, Mumford C, Jefferys G., Bains G. A study of evaporation from, and drying of, single droplets. In: Mujumdar AS. (editor) Proceedings of the 6th International Symposium in Drying IDS'88. Versailles, France.1988.

17. Minoshima H, Matsushima K, Liang H, Shinohara K. Estimation of Diameter of Granule Prepared by Spray Drying of Slurry with Fast and Easy Evaporation. Journal of Chemical Engineering of Japan 2002; 35, 880-885.

18. Seydel P. Experimental and Mathematical Modeling of Solid Formation at Spray Drying. Chemical Engineering Technology 2004; 27 (5): 505-510.

19. Perdana J, Fox M, Schutyser M, Boom R. Single-Droplet Experimentation on Spray Drying: Evaporation of a Sessile Droplet. Chemical Engineering & Technology 2011; 34 (7): 1151–1158.

20. Cheong H, Jeffreys G, Mumford C. A Receding Interface Model For The Drying of Slurry Droplets. AIChE Journal 1986; 32, 1334-1346.

21. Farid M. A New Approach to Modeling of Single Droplet Drying. Chemical Engineering Science 2003; 58, 2985-2993.

22. Almubarak A, Mumford C. Characteristics of the receding evaporation front in convective drying. In: Mujumdar AS. (editor) Proceedings of the 9th International Drying Symposium, IDS'94, Marcel Dekker Inc., NY, USA; 1994.

23. Kays W. M, Crawford M.E. Convective Heat and Mass Transfer, 2nd. ed., McGraw-Hill Book Company, New York, NY, USA.1980.

24. Thomas L. C. Heat Transfer professional Version. 2nd ed.: Capston Publishing Corporation, OK. USA. 1999.

25. Kreith F, Bohn M. Principles of Heat Transfer. 6th ed.: Brook/Cole Publishers, Pacific Grove, CA, USA 2001.

26. Almubarak A, Al-Saeedi J, Shoukry M. Effect of Boundary Layer on Mechanisms of Beach and Desert sand. European Journal of Soil Science 2008; 59, 807-816.

27. Waananen K, Litchfield J, Okos M. Classification of Drying Models for Porous Solids, Drying Technology 1993; 11, 1-40.

28. Holman J.P. Heat Transfer: McGraw-Hill Book Co., NY, USA 2002.

29. Nesic S, Vodnik J. Kinetics of Droplet Evaporation. Chemical Engineering Science 1991; 46, 527-537.

30. Chen X, Peng F. Modified Biot Number in The Context of Air Drying of Small Moist Porous Objects. Drying Technology 2005; 23, 83-103.

31. Hayder M, Mumford C. 1993. Mechanisms of Drying of Skin Forming Materials. Drying Technology 1993; 11, 1713-1750.

10

CHAPTER

Stochastic Assessment of Pillar Stability at Laisvall Mine Using Artificial Neural Network

Musa Adebayo Idris[a,*], David Saiang[a,b], Erling Nordlund[a]

[a] *Division of Mining and Geotechnical Engineering, Luleå University of Technology, SE-971 87 Luleå, Sweden*

[b] *SRK Consulting (Sweden) AB, SE-931 31 Skellefteå, Sweden*

HIGHLIGHTS

- We quantify the variability of the rock mass properties at Laisvall mine.
- We assess the probability of pillar failure at the mine.
- Finite difference code coupled with neural networks was used for the assessment.
- Increasing coefficient of variation of input parameters increases the pillar failure.
- Increasing mining depth and reducing pillar dimension increases the pillar failure.

ABSTRACT

Stability analyses of any excavations within the rock mass require reliable geotechnical input parameters such as in situ stress field, rock mass strength and deformation modulus. These parameters are intrinsically uncertain and their precise values are never known, hence, their variability must be properly accounted for in the stability analyses. Traditional deterministic approaches do not quantitatively consider these uncertainties in the input parameters. To incorporate these uncertainties stochastic approaches are generally used. In this study, a stochastic assessment of pillar stability using Artificial Neural Network (ANN) is presented. The uncertainty in the rock mass properties at the Laisvall mine were quantified and the probability density function of the deformation modulus of the rock mass was determined using probabilistic approach. The variability of the in situ stress was also considered. The random values of the deformation modulus and the horizontal in situ stresses were used as input parameters in the FLAC^{3D}numerical simulations to determine the axial strain in the pillar. ANN model was developed to approximate an implicit relationship between the deformation modulus, horizontal in situ stresses and the axial strain occurring in pillar due to mining activities. The closed-form relationship generated from the trained ANN model, together with the maximum strain that the pillar can withstand was used to assess the stability of the pillar in terms of reliability index and probability of failure. The results from this study indicate that, the thickness of the overburden and pillar dimension have a substantial effect on the probability of failure and reliability index. Also shown is the significant influence of coefficient of variation (COV) of the random variables on the pillar stability. The approach presented in this study can be used to determine the optimal pillar dimensions based on the minimum acceptable risk of pillar failure.

Keywords: Stochastic assessment; Probability of failure; Reliability index; Artificial Neural Network; Pillar stability

1. INTRODUCTION

A pillar can be defined as the in-situ rock mass between two or more underground openings. It is the main support in room and pillar mines. The support provided by the pillars controls the rock mass displacement throughout the zone of influence of mining, while the mining proceeds.

The analysis and design of mine pillars generally seek to optimize the size of the pillars so as to maximize the extraction ratio (i.e. amount of ore extracted relative to the total amount of ore available) while maintaining the stability of the mine. Hence the design of pillars has both economic and safety implications. The knowledge of the pillar strength and the determination of the required safety factor for a given loading condition are the most important aspects of pillar design. Conventional pillar design methods comprise the calculation of the mean pillar stress (e.g. the tributary area method and the method byCoates (1981) and the estimation of the pillar strength using empirical formulae (e.g.Obert and Duvall, 1967, Krauland and Söder, 1987 and Sjöberg, 1992). Based on the stress and strength of the pillar the factor of safety can be calculated. The factor of safety is the ratio of the pillar strength to the induced stress in the pillar and the pillar fails when the ratio is less than 1.

Though the conventional methods are widely used for pillar design, Alber and Heiland (2001) have expressed some concerns about this conventional approach for pillar design at shallow depth. They observed that the pillar failure at shallow depth could not be properly explained by comparing pillar strength with stresses induced on the pillar by mining activities. They suggested amongst other approaches that pillar failure could be related to strain. Therefore, when considering the strain occurring in the pillar the factor of safety can be determined as the ratio of the maximum strain that a pillar can withstand to the strain occurring in the pillar due to mining activities. Nevertheless, either ways of determining the factor of safety are largely deterministic and do not consider the inherent variability of the rock mass properties and that of the in situ stress field. Mean values of these input parameters are generally assumed. The results from the deterministic approach could be misleading depending on the distributive character of the rock property variation (Kim and Gao, 1995). Deng et al. (2003) have reported instances where pillars failed despite the fact that the failed pillars had been considered stable with factor of safety greater than 1.

Therefore, for a reliable design and analysis of construction elements such as mine pillars appropriate methods which incorporate the variability in the rock mass properties must be used. The methods which consider this variability are known as stochastic or probabilistic methods. With a stochastic approach, the stability analysis can be considered as a random system, where the occurrence of a pillar failure is a random event depending on the outcome of the random variables involved.

A number of stochastic approaches have been applied to various geotechnical problems, including underground excavation problems (e.g. Chen et al., 1997, Lilly and Li, 2000,Cai, 2011, Idris et al., 2011 and Dohyun et al., 2012), tunnel support (e.g. Schweiger et al., 2001, Li and Low, 2010 and Oreste, 2005), subsidence (e.g. Torano et al., 2000) and pillar stability (e.g., Pine, 1992, Joughin et al., 2000, Griffiths et al., 2002, Deng et al., 2003, Cauvin et al., 2009, Najafi et al., 2011, Recio-Gordo and Jimenez, 2012 and Wattimena et al., 2013). Pine (1992) presented a probabilistic approach for pillar design whereby normal probabilistic distributions were assumed for the random variables and the safety margin. Joughin et al. (2000) employed the point estimate method (Rosenblueth, 1981) to account for rock strength variability in the probabilistic method they presented for the design of chromite pillars in South Africa. Griffiths et al. (2002) analysed the stability of underground pillar by using random field theory with elasto-plastic finite element algorithm in a Monte Carlo framework. Deng et al. (2003)presented a probabilistic mine design method which combines the finite element methods, neural network and reliability analysis. Cauvin et al. (2009) used probabilistic approach to assess the effect of uncertainty in mining pillar stability analysis. Najafi et al. (2011) utilized First Order Second Moment (FORM) and Advanced Second Moment (ASM) for the probabilistic stability analysis of chain pillar in a coal mine in Iran. A probabilistic model based on linear classifier theory to predict the behaviour of pillar in longwall and retreat room and pillar mining was presented by Recio-Gordo and Jimenez (2012). Wattimena et al.

(2013) employed logistic regression to predict the probability of coal pillar stability for given pillar geometry and stress condition.

In general, stochastic assessment of pillar stability is performed by two procedures: the first step is to quantify the uncertainty in the rock mass properties in order to determine the basic statistical parameters (i.e. mean and variance) and probability density functions (PDFs) of the strength and deformation modulus of the rock mass using the Monte Carlo method. The Monte Carlo (MC) simulation technique is often adopted in the geotechnical stochastic analyses with implicit or explicit solutions but when the analysis is associated with numerical modelling then the MC simulation technique becomes time consuming and less appealing.

In the second step, the probability of failure is determined with respect to a specific failure criterion, which can either be the induced pillar stress exceeding the pillar strength or the strain occurring in the pillar exceeding the defined threshold strain value for the pillar. The onset of failure in the context of this study is defined as the limit state when the peak strength of the pillar is exceeded or the strain occurring in the pillar exceeds the peak strain for the pillar. For underground excavations this limit state is not known explicitly, instead numerical analysis using the finite difference method (FDM) or the finite element methods (FEM) can be combined with function approximation tools to construct a closed-form expression for the limit state surface. Recently, many function approximation tools have been proposed such as the response surface method (RSM), the point estimate method (PEM), and the Artificial Neural Network (ANN) to model the relationship. ANN, due to its high performance, has been one of the tools used in geotechnical engineering to model the relationship between non-linear multivariate variables (Sonmez et al., 2006).

In this study, a stochastic approach was used to analyse the pillar stability at the Laisvall mine in Sweden while considering the variability in the rock mass properties and in the in-situ stresses. The uncertainty in the rock mass properties at the Laisvall mine were quantified and the probability density function of the deformation modulus of the rock mass was determined. Also the variability of the horizontal in situ stresses was considered. The random values of these parameters (i.e. deformation modulus and horizontal in situ stresses) were used as input parameters for the FLAC^{3D} (Itasca, 2012) analyses to determine the axial strain in the pillar. The ANN model was developed to approximate an implicit relationship between the deformation modulus, horizontal stresses and the pillar axial strain within the range of possible values of the random input parameters. The closed-form relationship generated from the trained ANN model together with critical axial strain, which the pillar can withstand, was used to define a pillar performance function. The performance function was used to assess the stability of the pillar in terms of probability of failure and reliability index.

1.1. Artificial Neural Network (ANN)

The ANN, also referred to as neural network, is an information system that imitates the behaviour of the human brain by emulating the operation and connectivity of the brain to generate a general solution to a problem. ANN can be used to extract patterns and detect trends from problems where the relationship between the inputs and outputs are not sufficiently known. In recent years, ANN has been frequently used for functions approximation in different fields of science, including geotechnical engineering (Sahin et al., 2001). Basically, ANN consists of simple interconnected nodes or neurons as shown in Fig. 1 where *p* is the input, *w* is the weight, *b* is the bias, *f* is the transfer function and *a* is the output.

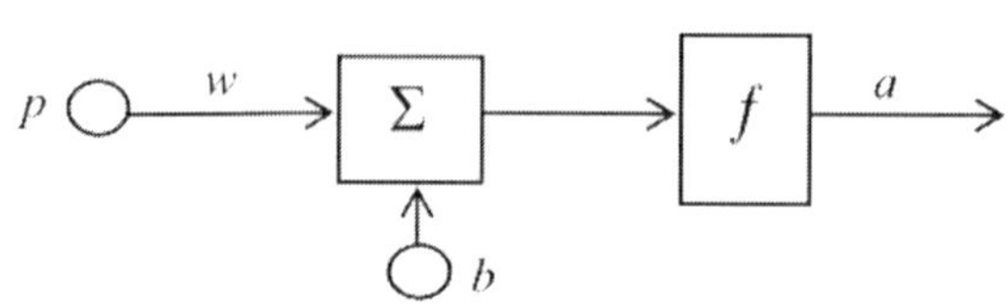

Fig. 1. Simple structure of ANN model.

If the neuron has N number of inputs then the output a can be calculated as:

$$a = f(\sum_{i=1}^{N} w_i p_i + b) \tag{1}$$

There are different types of transfer functions that can be used in ANN such as hard limit transfer function, linear transfer function, log-sigmoid transfer (Beale et al., 2012). The choice of the transfer function depends on the specification of a problem that the neuron is attempting to solve (Beale et al., 2012).

The architecture of ANN consists of the number of layers, the number of neurons in each layer and the neuron transfer functions. Two or more neurons can be combined in a layer and a network could contain one or two layers whereby each layer has different roles. There are output layers, input layers and intermediate layers or hidden layers. There is always one output layer and one input layer while there can be many hidden layers for an ANN. However, it is known that a network with one hidden layer can approximate any continuous function provided with sufficient connection weights (Hornik et al., 1989). The number of neurons in the input and output layer are determined by the number of the model input and output variables. There is no any exact guide for determining the number of neurons in the hidden layer, however some researchers (e.g. Aldrich and Reuter, 1994, Kanellopoulas and Wilkinson, 1997 and Seibi and Al-Alawi, 1997) have proposed heuristic relations for determining the neuron size.

Before an ANN can be used to make projections or predictions it has to be trained. In the training or learning process, the network is presented with a pair of training datasets including input and corresponding target values. The network computes its own output using its initial weights and biases. Then, the weights and biases are adjusted iteratively to reduce the errors between the network output and the target output. Mean square error is used as error index during the training phase to improve the network performance (Tawadrous et al., 2009). One of the most commonly used learning algorithms in geotechnical engineering is the back-propagation algorithm (BPA) proposed byRumelhart et al. (1986). In BPA there are two phases: forward prediction which calculates the output values of the ANN from training data and error back-propagation which adjusts the weight. There are many techniques to adjust the error but the steepest descent method is often used (Tawadrous et al., 2009). Once the training process of ANN is completed predictions can be made for a new input dataset.

2. DESCRIPTION OF THE CASE STUDIED

The Laisvall mine in northern Sweden, although closed in 2001, was a lead–zinc mine owned and operated by Boliden Mineral AB. The orebody, hosted in flat-lying quartzitic sandstone interlayered with clayey sandstones, was mined using the room and pillar mining method. The annual production before the closure of the mine was 1.6 Mt. There were four main ore bodies in the mine, most of them situated in the lower sandstone, except the Nadok orebody which was entirely situated in the upper sandstone. The thickness of the overlying strata above the Nadok orebody varied between 110 and 300 m (Söder and Krauland, 1990). In most of the orebodies of the mine the roof consisted of the same type of sandstone as the ore itself.

Söder and Krauland (1990) conducted a full-scale pillar test between 1983 and 1988 in the Nadok orebody. This test comprised 9 pillars and the objective was to determine the stress level in the pillar at failure (i.e., the bearing capacity) to serve as input to pillar design and long term planning of the mine. The pillars were subjected to increasing stresses by decreasing the cross-sectional area of the pillars through blasting of slices of approximately 0.4 m thickness, thereby reducing the widths and lengths of the pillars in each of the mining steps. The pillars were initially 4.6 m high, 7.4 m wide and 8.1 m long. Pillar failure appeared to have occurred after six slices were blasted.

3. STOCHASTIC ESTIMATION OF ROCK MASS PROPERTIES USING *GSI* SYSTEM

Inputs for the rock mass parameters in the numerical simulation were determined stochastically from the *GSI* (Geological Strength Index) rock mass characterization system. The *GSI* system, developed by Hoek et al. (1995), is based on the description of the rock structure and the block surface condition. It is a system that provides a complete set of mechanical properties for design purposes when used in the generalized Hoek–Brown criterion (Cai et al., 2007). The mechanical rock mass properties are the deformation modulus (E_m), the Hoek–Brown strength parameters (m_b, s and a), the tensile strength, and the equivalent shear strength parameters: cohesion (c) and the internal friction angle (ϕ). When the *GSI* system is used for rock mass characterization, two groups of parameters need to be determined, which includes the uniaxial compressive strength (UCS) and the material constant (m_i) of intact rock. The other group is the joint parameters which consist of joint geometry and strength parameters (Cai, 2011). These parameters varied considerably for the rock mass where the pillar test was conducted as reported by Söder and Krauland, 1990, Palmström, 1995 and Exadaktylos and Stavropoulou, 2008.

3.1. Evaluation of *GSI* from V_b and J_c

The determination of *GSI* based on visual observation may be subjective. Recently, Cai et al. (2004b) proposed a new approach to quantitatively determine *GSI* based on the block size and condition; block volume (V_b) and the joint condition factor (J_c). This approach is well suited for the stochastic approach and therefore adopted for this study.

The block size is determined from the joint spacing, joint orientation, number of joint sets and joint persistence. According to Cai et al. (2004b) the effect of the intersection angle between joint sets is relatively small compared to the joint spacing hence they suggested that for practical purposes the block volume for three or more joints can be approximated using

$$V_b = S_1 S_2 S_3 \tag{2}$$

where S_i is the spacing of each joint set.

The joint surface condition (J_c) is defined by the roughness, weathering and infilling. It is similar to the factor used by Palmström (1995) to quantify the joint surface condition. The combination of these factors defines the strength of a joint or block surface. Cai et al., 2004b defined J_c as:

$$J_c = \frac{J_W J_S}{J_A} \tag{3}$$

where J_W and J_S are the large-scale waviness and small-scale smoothness, respectively (Barton and Bandis, 1990 and Palmström, 1995) and J_A is the joint alteration factor (Barton et al., 1974).

Once the value of J_c and V_b are known the *GSI* value can be determined using the equation proposed by Cai and Kaiser (2006) which is expressed in terms of J_c and V_b, thus

$$\mathrm{GSI} = \frac{26.5 + 8.79 \ln J_c + 0.9 \ln V_b}{1 + 0.0151 \ln J_c - 0.0253 \ln V_b} \tag{4}$$

By extension, since the variability in the material properties is considered in this study, if the mean values and the coefficient of variations of J_c and V_b together with their probability density functions (PDF) are known then they could be used as input variables to determine the distribution of the *GSI* using the Monte Carlo simulation.

Palmström (1995) has conducted extensive evaluation of the joint characteristics of the pillars at the Laisvall mine and reported the V_b and J_c values for the rock mass at the mine. Exadaktylos and Stavropoulou (2008), based on the results of Palmström (1995), presented the range of values for the V_b and J_c, respectively, for the quarzitic sandstone pillar at the Laisvall mine. The mean values and the coefficient of variation (COV) for both V_b and J_c were estimated based on the "Three-sigma rule" described by Dai and Wang (1992). The three-sigma rule is based on the fact that 99.73% of all values of a normally distributed parameter fall within three standard deviations of the mean. The normal distribution is assumed for both V_b and J_c and the distributions were truncated at their maximum and minimum values. The mean values and the COV for the V_b and J_c together with their respective truncated PDF were used as inputs in Eq. (4) to generate the distribution of the *GSI* using the Monte Carlo simulation technique available in the Excel add in program @RISK (Palisade, 2001). Table 1 shows the statistical parameters for V_b and J_c and statistical parameters for the *GSI* generated from the Monte Carlo simulation.

Table 1. Statistical parameters for V_b, J_c and *GSI*.

Parameters	Min. value	Max. value	Mean	COV (%)	Distribution
J_c	0.75	2.25	1.5	16.7	Normal
V_b ($\times 10^3$ cm^3)	100	300	200	16.5	Normal
GSI	51	63.54	58.7	4.0	Normal

3.2. Determination of statistical parameters for UCS and, m_i

The *GSI* value, the UCS and the m_i of the intact rock are basic inputs for the generalized Hoek–Brown failure criterion (Hoek et al., 2002). The UCS and m_i of the intact rock are determined by laboratory testing as described by Hoek and Brown (1997). In this study, the mean value and the range for UCS used for the rock types in the Nadok orebody have been reported by Söder and Krauland, 1990 and Palmström, 1995. The information about the variation of the *UCS* is not reported. A normal distribution with a COV of 15% was assumed for the UCS. Generally, the COV of 15–20% is applicable to most natural geotechnical materials (Rethati, 1988). A normal distribution with the mean value of 17 and COV of 10% was assumed for the m_i. This was taken from the value suggested for sandstone by Marinos and Hoek (2000). Table 2 shows the statistical parameters for the UCS and m_i used in this study.

Table 2. Statistical parameters for the UCS and m_i.

Parameters	Mean value	COV (%)
UCS (MPa)	210	15
m_i	17	10

3.3. Determination of the rock mass properties and their Probability Density Functions

The generalized Hoek–Brown criterion for jointed rock masses was employed to estimate the rock mass strength parameters and deformation modulus for the Nadok orebody. At failure the generalized Hoek–Brown criterion relates the major effective principal stress σ_1 to the minor effective principal stress σ_3 as follows

$$\sigma_1 = \sigma_3 + \sigma_{ci}\left(m_b \frac{\sigma_3}{\sigma_{ci}} + s\right)^a \quad (5)$$

where m_b, s, a are Hoek–Brown strength parameters for the rock mass and σci is the UCS of the intact rock. Having determined the statistical parameters (i.e. mean and COV) of *GSI* and m_i Monte Carlo simulation technique was used to determine the distributions of the Hoek–Brown strength parameters for the orebody using the following equations:

$$m_b = m_i \exp\left(\frac{\text{GSI} - 100}{28 - 14D}\right) \quad (6)$$

$$s = \exp\left(\frac{\text{GSI} - 100}{9 - 3D}\right) \quad (7)$$

$$a = \frac{1}{2} + \frac{1}{6}\left(e^{-GSI/15} - e^{-20/3}\right) \tag{8}$$

where D is a factor which depends on the degree of disturbance the rock mass has been subjected to by blast damage and stress relaxation. It varies from 0 for undisturbed in situ rock masses to 1 for very disturbed rock masses. For this study, blasting damage was assumed to be minimal, hence $D = 0$ was used.

Different equations have been proposed to determine the deformation modulus of the rock mass based on the range of the UCS (Hoek et al., 2002). Hoek and Diederichs (2006) suggested that when reliable estimates of Young's modulus for intact rock are available, the deformation modulus of the rock mass (E_m) can be calculated from:

$$E_m = E_i\left(0.02 + \frac{1 - D/2}{1 + e^{((60+15D-GSI/11))}}\right) \tag{9}$$

where E_i is the Young modulus and it was determined from the modulus ratio (MR) proposed by Deere (1968) thus:

$$E_i = MR\sigma_{ci} \tag{10}$$

The value of MR was determined from the mean values of UCS and E_i for the intact rock reported by Söder and Krauland (1990). Eqs. (9) and (10) were used to estimate the statistical parameters and the PDF for deformation modulus (E_m) for the rock mass by using the already determined PDFs of the UCS and *GSI*. Similar to the previous parameters, the PDF and the statistical parameters were generated using the Monte Carlo simulation technique available in @RISK. Table 3 shows the probability density function (PDF), mean, standard deviation (STDEV) and the coefficient of variation (COV) for the deformation modulus (E_m).

Table 3. PDF, mean, standard deviation and coefficient of variation (COV) for E_m.

Parameter	Mean	STDEV	COV (%)	PDF
E_m (GPa)	23.47	4.04	17.17	Normal

4. VARIABILITY IN IN-SITU STRESS

The knowledge of in situ stresses is essential for the design of underground excavations and construction elements such as mine pillars. Due to spatial variation of the in-situ stress field at various locations within the rock mass, there is always an uncertainty in the results of in-situ stresses measurements (Cai, 2011).

In-situ stresses are usually reported as; vertical stress (σ_v), maximum horizontal stress (σ_H), and minimum horizontal stress (σ_h). The vertical stress is normally assumed to vary linearly with depth as:

$$\sigma_v = \rho g z \tag{11}$$

where ρ is the rock density, g is the gravity and z is the overburden. For this study, the rock density for the rock mass was assumed to be constant with a value of 2700 kg/m^{3}and the overburden varied from 110 to 300 m.

The maximum horizontal stress (σ_H) and minimum horizontal stress (σ_h) recorded at the mine varied between 20 MPa and 25 MPa (Marklund, 2013). Normal distribution was assumed for the horizontal in situ stresses with coefficient of variation (COV) of 3.7% using the three sigma rule. Only the variability in the horizontal stresses was considered in this study.

5. NUMERICAL ANALYSIS

Since the objective of the numerical analysis was to determine the induced axial stress and the elastic axial strain occurring in the pillar due to the mining, linear elastic analyses were considered sufficient.

Series of elastic numerical analyses were carried out using the FLAC[3D] code (Itasca, 2012) considering the possible variability in the rock mass properties and the in-situ stress conditions.

5.1. Model geometry

The layout of the test area and the geometry of the simulated pillars are shown in Fig. 2. Roller boundary conditions were used along all external boundaries. The mesh size used for the model was 0.4 m × 0.4 m × 0.4 m. It was chosen based on the results from the FLAC analysis where the induced axial stress in pillar was monitored for different mesh sizes until further reduction in the mesh size gave no significant difference in the model result.

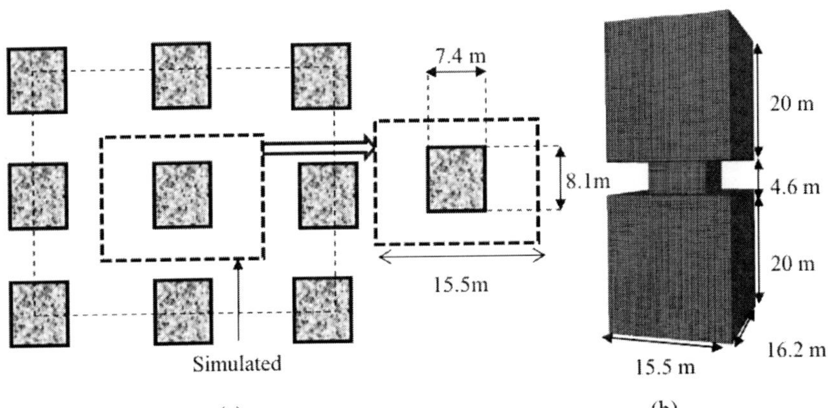

Fig. 2. The layout of the test area (a) and the geometry of the simulated pillar (b).

5.2. Modelling sequence

The modelling sequence used in this study consists of the following steps:

i. Generation of the model grid using a brick-shaped mesh.

ii. Selection of appropriate boundary conditions and material model.

iii. Initialization of in-situ stresses and material properties.

iv. Stepping of the model to initial equilibrium prior to any excavations.

v. Excavation of the roadways to create the rib pillar and later excavating the rooms across the rib pillar to create the panel pillar. These excavations were henceforth referred to as mining step 0 in the modelling sequence.

vi. Sequential loading of the pillar induced by blasting the pillar in 6 steps, henceforth referred to as mining steps 1–6. Each step extracts one slice of approximately 0.4 m from the length of the pillar and one slice of approximately 0.4 m from the width of the pillar.

Fig. 3 shows the mining steps 1–6 and Table 4 shows the pillar dimensions corresponding to the mining steps. At each mining step the axial stress and maximum vertical displacement at both ends of the pillar were obtained. The mean pillar stress was estimated by averaging the vertical stress component computed by FLAC3D along the center line *ab* (see Fig. 4) of the pillar. The axial strain (ε) was calculated as follows:

$$\varepsilon = \frac{U_b - U_a}{H} \times 100 \tag{12}$$

where U_a and U_b are the vertical displacement at points a and b, respectively while H is the height of the pillar.

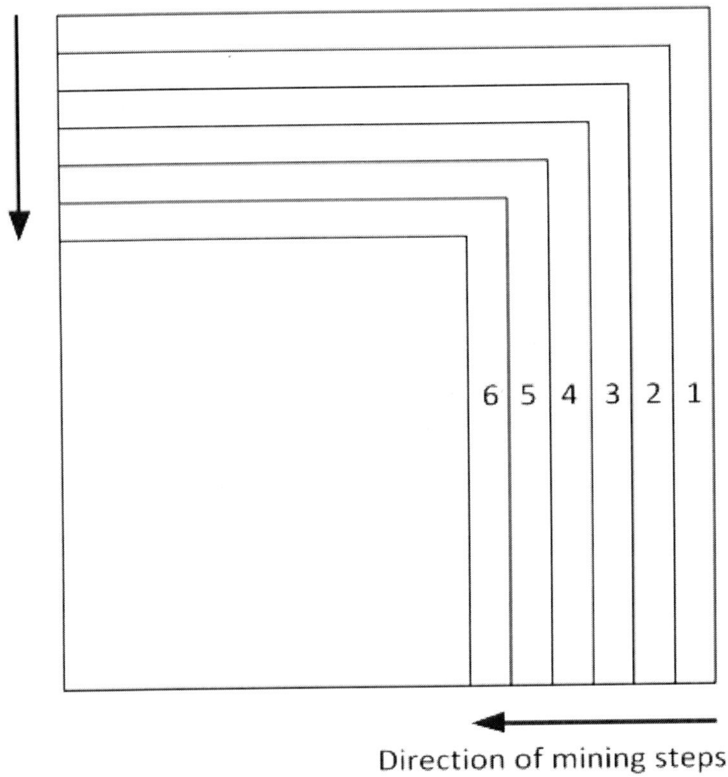

Fig. 3. Plan view of the simulated pillar showing the mining steps (not to scale).

Table 4. Mining steps and corresponding pillar dimension.

Mining steps	Height, *H* (m)	Width, *W* (m)	Length (m)	*W*/*H* (approx.)
0	4.6	7.4	8.1	1.61
1	4.6	6.7	7.8	1.46
2	4.6	6.3	7.2	1.37
3	4.6	5.9	6.6	1.28
4	4.6	5.5	6.2	1.20
5	4.6	4.9	5.9	1.07
6	4.6	4.5	5.3	0.98

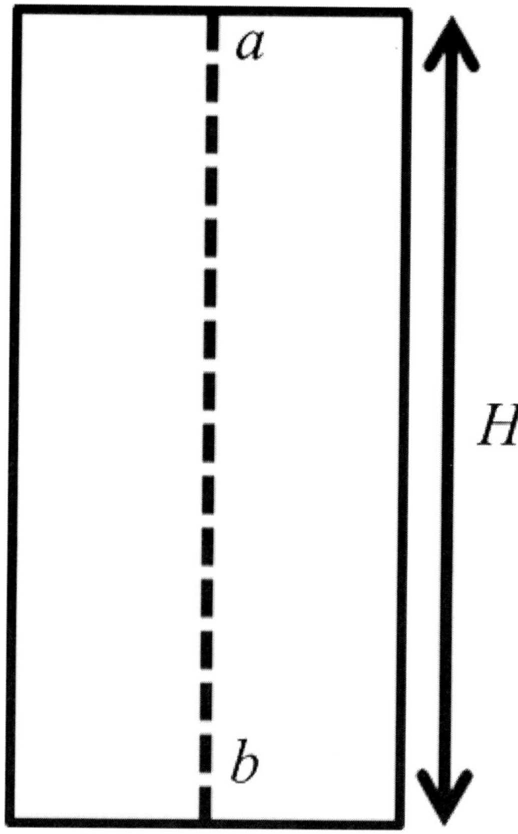

Fig. 4. Section of the pillar geometry showing the center line *ab* and height *H* (not to scale).

5.3. Sensitivity analysis

In the room and pillar mining method, roadways are usually developed ahead of mining and are initially subjected to the in-situ stress state. As mining progresses, the pillars are subjected to an increased induced stress which consequently leads to increase in the rock mass response in terms of displacement or strain. Traditionally, the stability of a pillar is assessed by using the criteria which compares the strength of the pillar to the maximum average stress in the pillar. The maximum average stress can be determined from two or three dimensional elastic numerical analyses while the strength is determined from empirical equations. The pillar stability can also be assessed by comparing the strain monitored in the pillar during the numerical analyses with the maximum strain that the pillar can withstand. Some of the input parameters for the elastic numerical analyses are the deformation modulus of the rock mass and horizontal in-situ stress. As discussed previously, these parameters exhibit inherent variability and uncertainty hence for this study there is need to find the appropriate criteria for assessing the stability of the pillar while considering the variability of these parameters. A sensitivity analysis was performed to determine the effect of the variability of these input parameters on the axial strain of and the mean axial stress in the pillar. Two cases were considered for the sensitivity analysis as follows:

i. The deformation modulus was varied between 20 and 32 GPa while the other input parameters were kept constant and equal to their mean values.

ii. The horizontal stresses were varied between 20 and 26 MPa while the other input parameters were kept constant and equal to their mean values.

The mean values of the input parameters used for the sensitivity analysis are shown inTable 5.

Table 5. Mean values of the input parameters for the sensitivity analysis.

Parameters	Mean values
Deformation modulus E_m, GPa	23.47
Maximum horizontal stress, σ_H, MPa	22.50
Maximum horizontal stress, σ_h, MPa	22.50
Vertical stress, σ_v, MPa	5.43

5.3.1. Results of the sensitivity analysis

The results of the sensitivity analysis are shown in Fig. 5 and Fig. 6. It is evident from Fig. 5 that the variation in the horizontal stress has significant effect on both axial strain and mean axial stress in the pillar. The strain and the mean stress increased when the width-to-height ratio of the pillar was reduced from 1.61 to 1.46 (i.e. mining step 0 to step 1). However, as shown in Fig. 6a the axial strain in the pillar decreases for an increase in the deformation modulus. Also the strain increases with reduction in the width-to-height ratio. Therefore, from Fig. 6a it is evident that the axial strain in the pillar is sensitive to the variations in the deformation modulus as well as in the pillar geometry. Conversely, the mean axial stress is not sensitive to any increase in the deformation modulus as shown inFig. 6a. However the axial stress increased when the width-to-height ratio was reduced (i.e. from mining step 0 to step 1). These results show that the distribution of elastic stress in the pillar is a function of the pillar geometry and the in-situ stress. This has also been observed by Martin and Maybee, 2000 and Lunder and Pakalnis, 1997.

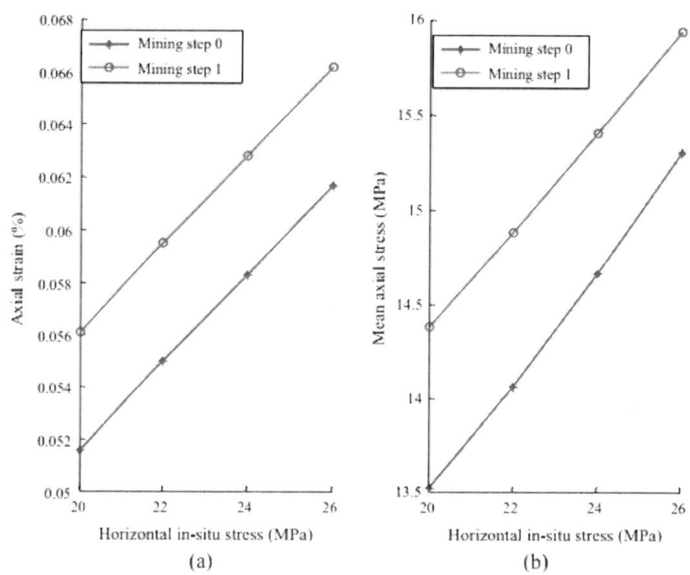

Fig. 5. Effect of varying the horizontal stresses on: (a) axial strain and (b) mean axial stress in the pillar at mining steps 0 and 1.

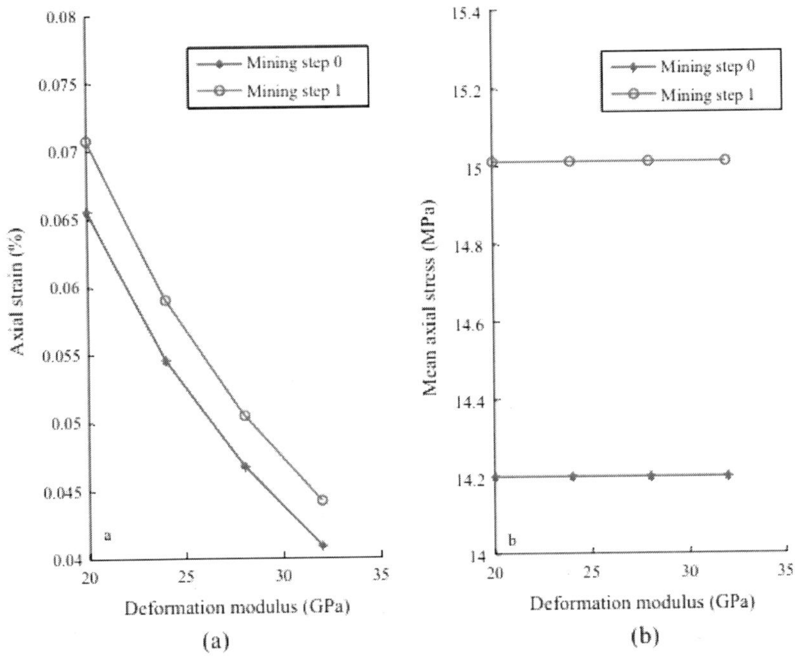

Fig. 6. Effect of varying deformation modulus on: (a) axial strain and (b) mean axial stress in the pillar at mining steps 0 and 1.

The sensitivity analysis shows that given the same pillar geometry the stress in the pillar, under elastic conditions, will be the same for pillars with different rock mass properties if they are subjected to the same in situ stresses. Deformation (e.g. strain) is the response of the rock mass when subjected to stress/load hence this response (i.e. strain) can be used to analyse the stability of the pillar instead of using a stress/strength criterion. Therefore, in this study strain-based criterion is used for assessing the pillar stability in this study.

5.4. Determination of the pillar performance function

In order to determine the probability of failure for the pillar as well as its reliability the relationship between the capacity and the demand on the pillar must be established. Concluding from the results of the sensitivity analysis presented above, the strain based criterion is adopted for this study. Therefore, the pillar capacity is expressed in terms of the critical strain ($\varepsilon_{critical}$) and the demand is the axial strain (ε).

Sakurai, 1981, Sakurai, 1997 and Sakurai, 1999 has developed a concept of critical strain for rock masses. The critical strain is the ratio of uniaxial compressive strength to Young's modulus i.e.

$$\varepsilon_{\text{critical}} = \frac{\sigma_{cm}}{E_m} \times 100 \qquad (13)$$

where $\varepsilon_{critical}$ is the critical strain, σ_{cm} is the uniaxial compressive strength (UCS) of the rock mass and E_m is the deformation modulus of the rock mass. The UCS of the rock mass σcm is obtained by setting $\sigma_3 = 0$ and $\sigma_1 = \sigma cm$ in Eq. (5), giving:

$$\sigma_{cm} = \sigma_{ci} s^a \qquad (14)$$

The rock mass constants *a* and *s* are obtained from Eqs. (7) and (8), respectively.

Since the critical strain may not be the strain at failure for most rocks, Sakurai (1981)proposed a relationship between strain at failure and critical strain as

$$\varepsilon_{\text{failure}} = \frac{\varepsilon_{\text{critical}}}{1 - R_f} \qquad (15)$$

where $\varepsilon_{failure}$ is the failure strain and R_f is a parameter representing failure strength. According to Sakurai (1981)R_f ranges from 0.05 to 0.8. The range of R_f suggested by Sakurai does not have a general application as it includes both weak and hard rocks. However, based on the study conducted by Cai et al. (2004a) on the generalized crack initiation and propagation thresholds of brittle rock masses Cai (2011) proposed the R_f ranging from 0.1 to 0.3 for hard rock masses. For this study 0.2 was assumed to be an approximated value for R_f for the rock mass at the Laisvall mine based on a rough estimate using the available data from the mine.

Therefore, for this study the performance function for the pillar is the limit-state function which defines the relationship between the failure strain and the axial strain due to mining. The limit state function, $g(X)$, and probability of failure P_f can be defined as:

$$g(X) = R(X) - S(X) \tag{16}$$

$$P_f = P[g(X) < 0] \tag{17}$$

where R is the failure strain and S is the axial strain. Both $R(X)$ and $S(X)$ are PDFs of the random variable X.

Aside from the limit state function, the safety index or reliability index (β) is another important quantity for the reliability of a pillar. The reliability index can be defined as:

$$\beta = \frac{\mu_m}{\sigma_m} \tag{18}$$

where μ_m and σ_m are the mean and standard deviation of the PDF of $g(X)$, respectively. The larger the reliability index, the smaller the probability of pillar failure. The limit state function, $g(X)$, is assumed to follow a normal distribution hence β can be defined as

$$\beta = -\phi^{-1}(P_f) = \phi^{-1}(1 - P_f) \tag{19}$$

where ϕ^{-1} is the inverse cumulative density function (CDF) of the standard normal variable at the probability level $(1 - P_f)$.

5.4.1. Determination of closed-form equation for axial strain

In this study an approach that integrates ANN and finite difference analysis was adopted to obtain a closed-form equation of the axial strain for each of the mining steps. The proceeding sections describe the processes involved in achieving this.

5.4.1.1. Determination of model training dataset

An important step in developing ANN models is to select the model input variables that have the most significant effect on the model performance (Faraway and Chatfield, 1998). For this study, the training datasets are the deformation modulus and horizontal in-situ stresses as input data and the corresponding axial strain as output or target data. In order to approximate the relationship between the axial strain and the random variables a training dataset is designed. Thirty-six (i.e. 6^2) combinations of random values of E_m and σ_H were generated by considering equally spaced six sampling points of each random values within the range of ±3 standard deviations of their respective means. This represents almost the total population of their normal distributions.

The FLAC[3D] elastic simulations were carried out using each of the 36 combinations of E_m and σ_H together with constant values for Poisson's ratio (0.2) and mean value for overburden (i.e. 205 m) to generate the corresponding 36 axial strains. The 36 datasets were divided into 3 sections randomly for the ANN model training, testing and validation. 70% of the datasets for the training, 15% for the testing and 15% for the validation.

The input and output data were normalized or scaled to be in the range of ±1 because the ANN algorithm especially the back-propagation, works best when the training datasets are scaled or normalized (Beale et al., 2012). The following equation was used for linear scaling of the input and target data (Goh and Kulhawy, 2003).

$$x_{norm} = 2\left(\frac{x - x_{min}}{x_{max} - x_{min}}\right) - 1 \qquad (20)$$

where x_{norm} is the normalized value of parameter x with maximum and minimum values of x_{max} and x_{min}, respectively. The ranges of values used for E_m and σ_H as input parameters for the training datasets were 23.47 ± 4.04 GPa and 22.50 ± 0.83 MPa, respectively. The corresponding maximum and minimum values of the output values of the axial strain for the base case (i.e. at depth of 205 m) are shown in Table 6.

Table 6. Maximum and minimum values of axial strain.

Mining step	Axial strain (%)	
	Max.	Min.
0	0.083	0.037
1	0.090	0.040
2	0.095	0.042
3	0.100	0.044
4	0.102	0.045
5	0.111	0.049
6	0.116	0.051

5.4.1.2. Designing the architecture of the ANN

The multilayer perceptron (MLP) feed-forward neural network was considered for this study. MLP is commonly used in geotechnical engineering (Sahin et al., 2001). For the development of the network, a commercial software package MATLAB (MathWorks Inc., 2012) was used to simulate the ANN operations. The network has one input layer, one output layer and one hidden layer. The input layer has two neurons and the output layer has seven neurons. The number of neurons in the hidden layer was chosen to be 5 utilizing a trial and error approach. The best configuration was achieved in this work by using the linear transfer function (purelin) in the output layer and the tansig transfer function in the hidden layer. The network was trained with Levenberg–Marquardt back-propagation algorithm (trainlm).

5.4.1.3. Model training, testing and validation

The process of optimizing the connection weights is known as training (Sahin et al., 2001). The weights and biases were initialized to non-zero random values. Then the normalized training dataset of the input and output values generated from the elastic simulations with FLAC^{3D} were presented to the network. The performance of the ANN model was measured in terms of an error criterion between the target output and the calculated output. The output calculated at the end of each feed-forward computation was compared with the target output to estimate the mean-squared error (MSE). Thereafter the back-propagation algorithm adjusted the weights and biases until the mean squared error was greatly minimized.

Fig. 7 shows the relationship between the output targets (axial strain values from FLAC^{3D}simulations) and the predicted values obtained from the ANN training, testing and validation process for the base case. Each diagram in the figure represents plots for different percentages of the total dataset, for training (70%), validation (15%), test (15%) and all (100%). The model shows good correlation to the training, validation, test and of course all the datasets with R values greater than 0.9. R values is an indication of the relationship between the outputs and targets where $R = 1$ indicates exact linear relationship.

5.4.1.4. Implementation of the trained network

Having trained the ANN by optimizing the weights and biases, the network can be used to predict target values given the input pattern within the range of the input values used for the network training.

The optimal trained connection weights and biases were extracted from the network model and used to develop a mathematical expression relating the input values, E_m andσ_H and the output variables, axial strain (ε), for each mining step. The network model has two input neurons, five hidden neurons, and seven output neurons, each one for each mining step hence the closed-form equation relating the inputs variables and output variables is as follows:

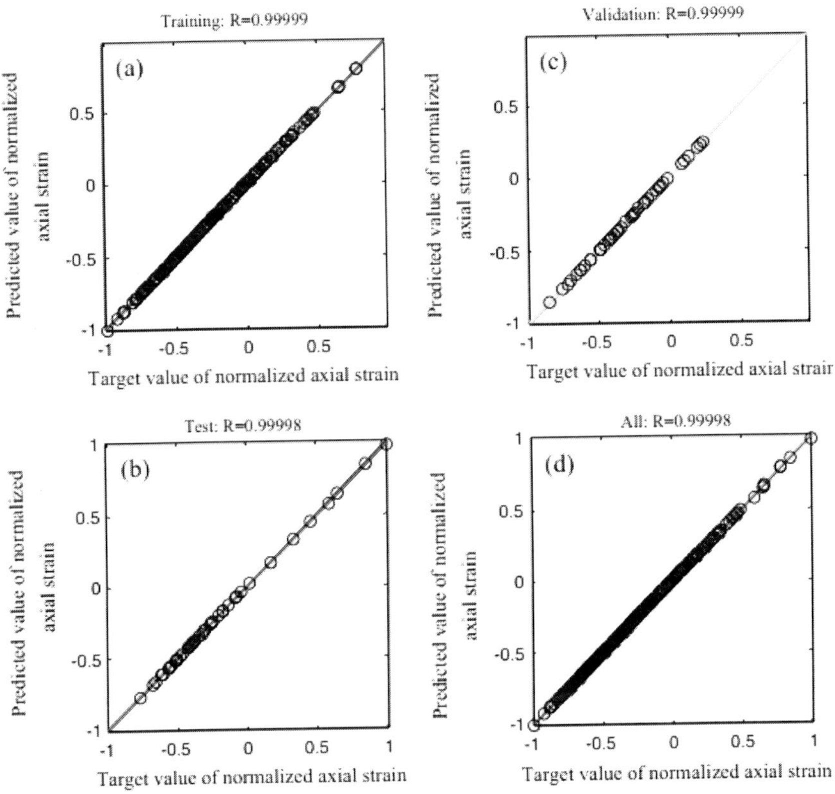

Fig. 7. Comparison of the predicted values of the normalized axial strain using ANN and the target values (obtained from the FLAC^{3D} simulations) for the base case: (a) training dataset (70% of the total dataset); (b) test dataset (15% of the total dataset); (c) validation dataset (15% of the total dataset); and (d) all the data sets.

$$T_n = f_2\{B_2 + LW(f_1(B_1 + IW.P_n))\} \tag{21}$$

where B_1 is the bias vector for the input, IW is the vector for weight connection between neurons of the hidden layer and the single output, B_2 is the vector for the bias at hidden layer neuron, LW is the matrix for the hidden layer weight. P_n is a matrix of normalized input vectors used as input values for the network training. T_n is the corresponding matrix of normalized output vectors and f_1 and f_2 are the transfer functions which are tansig and purelin, respectively. The normalized output vectors (T_n) could be converted to non-normalized output vectors (T) as follows:

$$T = 0.5(T_n + 1)(\max T - \min T) + \min T \tag{22}$$

where $\max T$ and $\min T$ are the vectors containing the maximum and minimum values of ε, respectively which were used for the network training and testing. These values for the base case have been presented in Table 6.

5.4.2. Pillar performance function

Using Eqs. (21) and (22), the pillar performance function can be expressed as:

$$g(X) = \varepsilon_{\text{failure}} - T. \tag{23}$$

The PDFs and the statistical parameters of σ_{ci}, s, a and E_m were used as input values to determine the random values of the critical strain ($\varepsilon_{\text{critical}}$) as shown in Eqs.(13) and (14). The strain at failure, which is the peak strain, was then determined using Eq. (15). The PDF and the statistical parameters of E_m and σ_H were used as the input values to generated T (i.e.ε) in Eq. (22). The Monte Carlo (MC) technique was used to simulate the performance function (i.e. Eq. (23)) using the program @RISK. In each simulation, the values of the variables X (i.e. σ_{ci}, s, a, E_m and σ_H) were randomly generated according to their PDFs. The PDFs, means and standard deviations of the performance function ($g(X)$) were determined from the MC simulation after 10^5simulations at each mining step. The probability of failure P_f for the pillar at each mining step was calculated as

$$P_f = P[g(X) < 0] = \frac{N_f}{N} \tag{24}$$

where N_f is the number of simulations with $g(x) < 0$ (i.e. number of pillar failures), and$N$ is the total number of simulations. The reliability index (β) was also estimated from the statistical parameters of the performance function using Eq. (18).

6. RESULTS AND DISCUSSIONS

In this section the results of the stochastic assessment of the pillar behaviour at the Laisvall mine are presented and discussed below.

6.1. Effect of mining activities on the pillar stability

The effect of the mining steps or the gradual reduction of the pillar width-to-height ratio on the stability of the pillar is demonstrated with the base case (i.e. at depth 205 m) and presented in Fig. 8. It is obvious from the figure that the mining steps have significant effect on the pillar stability. By reducing the cross-sectional area of the pillar the pillar strength is reduced thereby allowing more strain to occur in the pillar. Consequently the probability of pillar failure significantly increased as shown in Fig. 8, while the pillar reliability index decreased as the width-to-height ratio reduced.

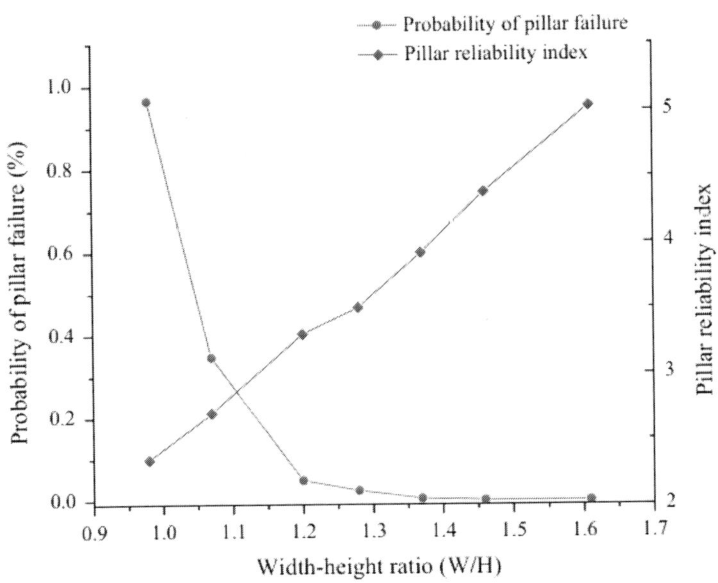

Fig. 8. Effect of mining activities on the pillar stability in terms of failure probability and reliability index (base case).

6.2. Effect of mining depth

As mentioned previously, the overburden varied between 110 and 300 m. In order to determine the effect of the various mining depths additional probabilistic analyses following the same procedure for the base case were performed. The mining depths considered were 220, 240, 260 and 280 m. Fig. 9 shows the effect of the mining depth on the performance of the pillar. It can be seen that the probability of pillar failure increases as the depth increases. Likewise, the corresponding pillar reliability index decreases as the mining depth increases; the lower the reliability index, the larger the corresponding probability of pillar failure. The failure probabilities of the pillar at all the considered depths are very small up to step 4 but increased significantly after step 4 for all the depths.

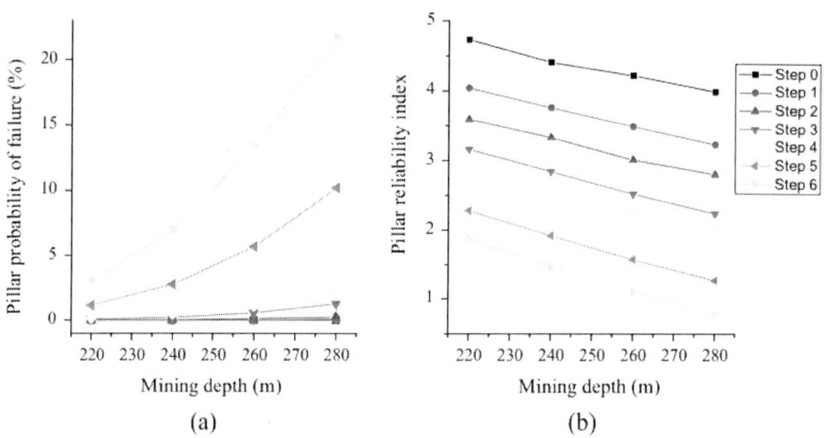

Fig. 9. Probability of failure (a) and reliability index (b) for different mining depths at different mining steps.

6.3. Effect of changing the coefficient of variation (COV) of the random variables on pillar stability

The UCS of the intact rock is a factor in the determination of the deformation modulus (E_m) and the critical strain as indicated in Eqs. (10) and (14). Therefore, any changes in the UCS will have effect on the pillar stability. The effect of changing the coefficient of variation (COV) of the UCS on the pillar stability was investigated individually using five different values of COV (10%, 15%, 20% 25% and 30%). The results for the pillar at a depth of 280 m are presented in Fig. 10 to illustrate the effect. The depth 280 m is chosen as its results contain both small and large values of the probability of failure hence it can represent the trend for the remaining depths.

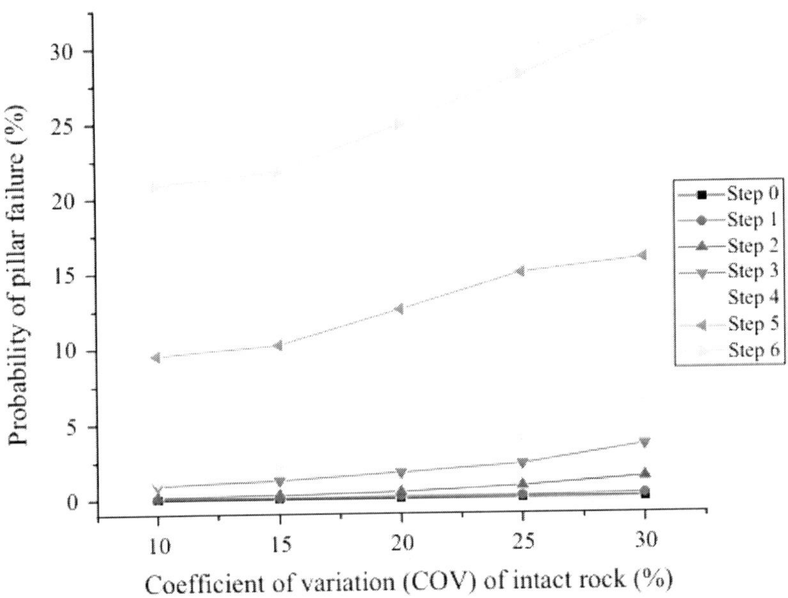

Fig. 10. Effect of varying COV of intact rock UCS on the pillar probability of failure at different mining steps.

Since COV is a non-dimensional measure of variability the smaller its value is the smaller the amount of uncertainty of the variable. It can be observed from Fig. 10 that the probability of failure increases as the COV of the UCS increases which shows that when the variability in the material properties of the rock mass increased then the probability of pillar failure will also increase. The variation in the COV has no effect on the mean factor of safety (Fig. 11). This shows that when the stability of a pillar is analysed in terms of reliability and probability of failure the conventional factor of safety is an inappropriate indicator of pillar stability. Therefore, in such a case the probability of failure or reliability index is the most useful criterion of pillar stability, because it combines information on both the mean values and the variability. Nevertheless, the factor of safety should not be abandoned in favour of reliability analyses but it has to be used as a complementary tool to reliability analyses (Dai and Wang, 1992).

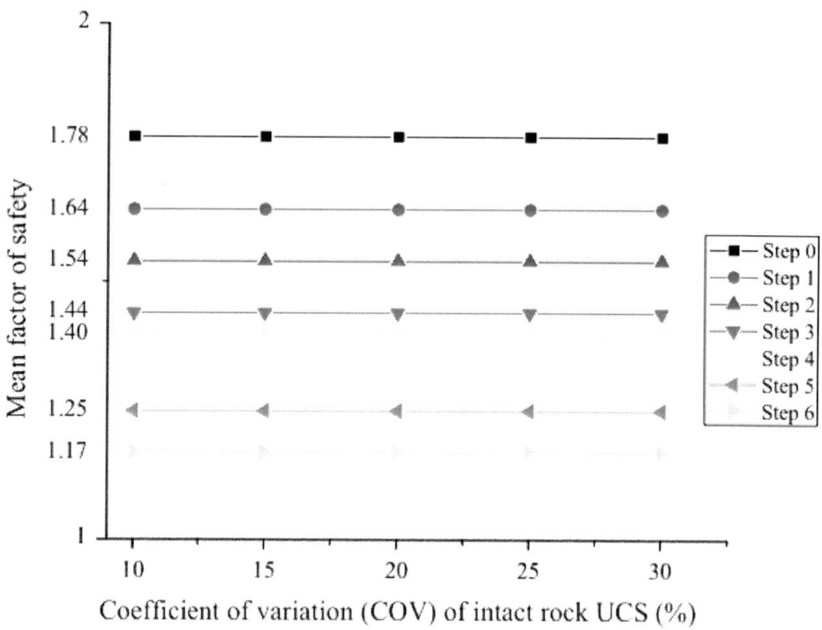

Fig. 11. Effect of varying COV of intact rock UCS on the mean factor of safety at different mining steps.

6.4. Assessment of pillar performance

The performance of a structure can be described in terms of the specific limit state being reached. The limit state is the acceptance criteria established by considering the nature, importance, and consequences of failure of the structure under consideration. Many performance criteria in terms of reliability index (β), probability of failure (P_f) and mean factor of safety (mFS) have been proposed for geotechnical structures (Lunder and Pakalnis, 1997, US Army Corps of Engineer, 1997 and Priest and Brown, 1983). The pillar performance for the different mining steps at different depth is hereby assessed based on the criterion proposed by US Army Corps of Engineer (1997). The criterion is summarized in Table 7. Fig. 12a–g show β and the corresponding performance levels for different depths and each mining step. The performance levels based on the US Army Corps of Engineer criterion are indicated in Fig. 12a–g. Targeted performance level of any geotechnical project depends on the nature of the project and the level of risk that can be accepted. Risk is a product of the probability of failure and the consequence of the failure. For instance an underground nuclear waste disposal will require high performance level because of the severity of the consequence of its failure. Priest and Brown (1983) proposed $\beta = 2.3$ for geotechnical structures if the consequences of failure are moderately serious. Hence for this study a performance level below average was chosen i.e. $\beta = 2.5$. Therefore, based on the chosen performance level for the pillar it can be seen from Fig. 12a–g that the targeted performance level, below average, can be achieved for the pillar at all depths at mining steps 0–2 (i.e. $W/H = 1.61–1.37$). The same performance level can also be achieved for mining depths 205, 220, and 240 m at mining step 4 (i.e. $W/H = 1.20$). At mining depth 205 m the targeted performance level can only be achieved at mining step 5. The performance level cannot be achieved for the pillar for all the depths at mining step 6.

Table 7. Typical values for reliability index and probability of failure and corresponding performance level.

Reliability index 4	Probability of failure	Performance level
5.0	2.9E−7	High
4.0	3.2E−5	Good
3.0	1.35E−3	Above average
2.5	6.21E−3	Below average
2.0	0.023	Poor
1.5	0.067	Unsatisfactory
1.0	0.159	Hazardous

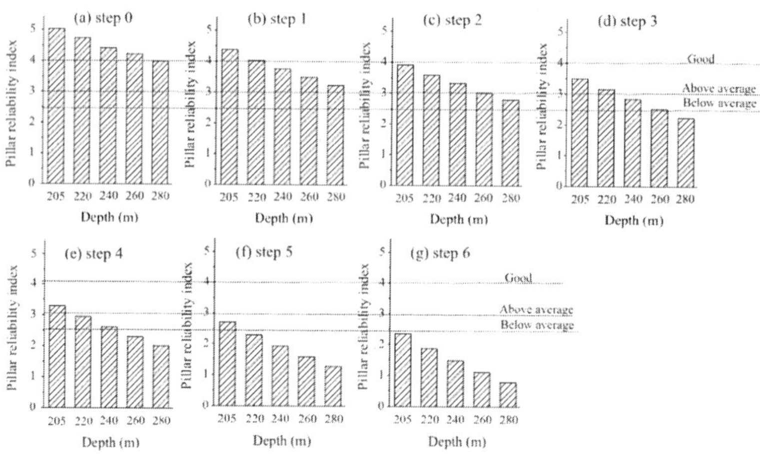

Fig. 12. Pillar reliability index (β) and the corresponding performance level for different depths at each mining step.

It is obvious that the optimum dimension of the pillar that can be achieved depends on the level of the risk that could be tolerated. Therefore, when a targeted performance level is chosen for an underground structure the procedure described in this paper can be used for optimal design of pillar dimension especially when the variability and uncertainty of the design input parameters are to be considered. However, when using this approach with the strain based criterion it should be noted that the value of R_f (failure strength parameter) has a significant influence on the probability of failure as well as the reliability index as illustrated in Fig. 13. Fig. 13 shows the pillar reliability index at mining depth 280 m for different R_f values. For $R_f = 0.1$ the performance level (i.e. below average) can only be achieved at mining step 0 while the same performance level can be achieved up to mining step 4 if R_f value is 0.3. Hence, the R_f value should be carefully determined through field monitoring programs for specific mine before using this approach.

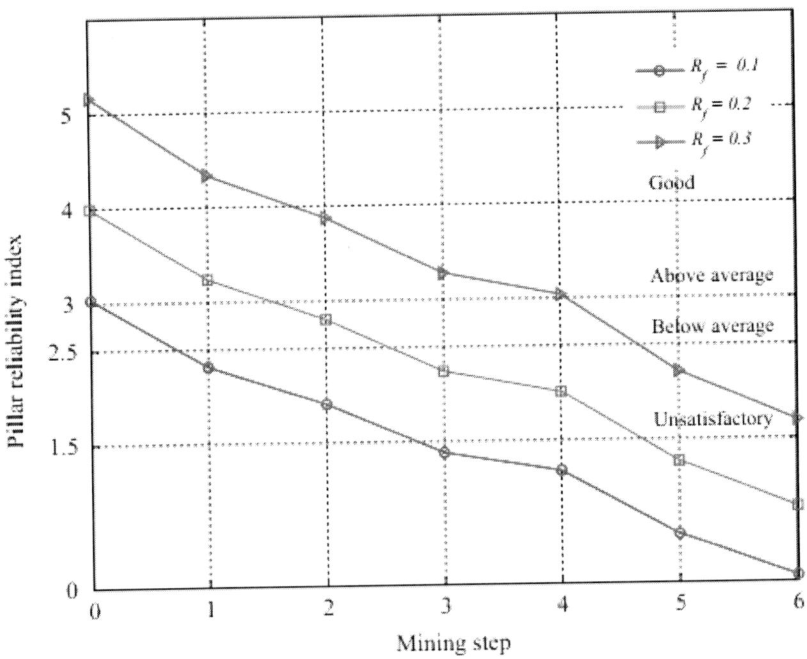

Fig. 13. Influence of R_f value on the pillar reliability index and performance level.

7. CONCLUSION

A stochastic assessment of the pillar stability of the Laisvall mine has been presented in this paper. The uncertainty and variability in the rock mass parameters as well as that of the horizontal in-situ stresses were considered. Elastic 3D numerical analyses were performed to determine the axial strain in the pillar at different mining steps for different depths. Artificial Neural Network (ANN) was used to generate a closed-form expression for the relationship between the random variables (i.e. E_m and σ_H) and the axial strain using the results of the 3D numerical analyses. The closed-form equation together with the strain at failure was used to evaluate the performance function. The performance function was determined from Monte Carlo simulations of 10^5 trials and the probability of failure and reliability index were determined from the simulation results. Based on the results and discussions presented in this study, the main conclusions are summarized as follows:

1. Rock mass properties as well as in-situ stresses are intrinsically variable s uch that using their mean values could have significant impact on the design performance as shown in this study. Therefore, this variability should be properly considered in the determination of the input parameters for the numerical analyses using a stochastic approach. This will enable the evaluation of the probability of failure during the planning stage, thereby reducing the risk to an acceptable level.

2. The strain-based criterion used in this study seems appropriate when the variability in the rock mass properties is considered in the analysis as pillars with different rock mass properties deform differently when subjected to the same stress condition. The strain-based criterion can also be used without having to determine the dimension of the pillar which is required to determine the strength of the pillar from empirical equation when using stress/strength criterion. Furthermore, strain is relatively cheaper and easier to measure in the field than the stress which is expensive and time consuming. Therefore using the approach presented in this study may offer some potential and advantages than the traditional strength/stress based criterion.

3. The trained ANN model is a good approximation tool to generate an implicit relationship between the random variables (E_m and σ_H) and the axial strain as shown in this

study, thereby reducing the number of $FLAC^{3D}$ numerical simulations. Also the ANN model was able to capture the tails of the distribution of the random variable because, the range of the value for the random variable which covers almost 99.7% (i.e. ±3 standard deviation from the mean) of the distribution was considered. It should be noted that the trained ANN model can only be used in the similar situation described in this study within the range of values of the training dataset.

4. The width-to-height ratio of a pillar is very significant to the stability of the pillar and as the ratio reduces through the mining steps the reliability index reduces and consequently the probability of failure of the pillar increases.

5. Increase in the in-situ stresses which could be linked to the increase in the depth affects the stability of the pillar. As the depth increases the probability of pillar failure increases and the reliability index of the pillar reduces.

6. The variation in the COV of the intact rock UCS has effect on the reliability index and the probability of failure. This study indicates that, an increase in the COV of the intact rock UCS results in an increase in the probability of failure and reduces the reliability index. However, the variation in the COV has no effect on the mean factor of safety. Therefore, the mean factor of safety may not be an adequate parameter to assess the stability of pillar when considering the variability and uncertainties in the rock mass material properties. Instead, the reliability index and probability of failure are the most useful indicators for assessing the stability of pillar in such situation.

7. The assessment of the pillar performance depends on the nature, importance and the consequence of the failure. The management decision on the level of risk that can be taken may affect the design of the pillar with regard to its dimension during the planning stage.

8. The approach presented in this study shows that when considering the uncertainty and variability of rock mass properties, the stochastic approach can be a useful tool for analysing and assessing the pillar stability. This will allow setting acceptable criteria for pillar design based on the minimum acceptable risk or performance level.

ACKNOWLEDGEMENTS

The authors acknowledged the Centre of Advanced Mining and Metallurgy (CAMM) at the Luleå University of Technology, Luleå, Sweden for financial support. The authors thank Mr. Per-Ivar Marklund of Boliden Mineral AB for providing information about the Laisvall mine.

REFERENCES

1. M. Alber, J. Heiland **Investigation of a limestone pillar failure. Part 2: Stress history and application of fracture mechanics approach** Rock Meck. Rock Engng., 34 (2001), pp. 187–199

2. C. Aldrich, M.A. Reuter **The application of neural nets in the metallurgical industry** Miner. Eng., 7 (1994), pp. 793–809

3. Barton, N.R., Bandis, S.C., 1990. Review of predictive capability of JRC–JCS model in engineering practice. In: Proceedings of International Symposium on Rock Joints, Balkema, Rotterdam, pp. 603–610.

4. N.R. Barton, R. Lien, J. Lunde **Engineering classification of rock masses for the design of tunnel support** Rock Mech., 6 (1974), pp. 189–239

5. M.H. Beale, M.T. Hagan, H.B. Demuth **Neural Network Toolbox User's Guide** The MathWorks Inc., Natick, MA (2012)

6. **Rock mass characterization and rock property variability considerations for tunnel and caven design** Rock Mech. Rock Eng., 44 (2011), pp. 379–399

7. M. Cai, P.K. Kaiser **Visualization of rock mass classification systems** Geotech. Geol. Eng., 24 (2006), pp. 1089–1102

8. M. Cai, P.K. Kaiser, Y. Tasaka, T. Maejima, H. Morioka, M. Minami **Generalized crack initiation and crack damage stress thresholds of brittle rock masses near underground excavations** Int. J. Rock Mech. Min. Sci., 41 (2004), pp. 833–847

9. M. Cai, P.K. Kaiser, H. Uno, Y. Tasaka, M. Minami **Estimation of rock mass strength and deformation modulus of jointed hard rock masses using the *GSI* system** Int. J. Rock Mech. Min. Sci., 41 (2004), pp. 3–19

10. M. Cai, P.K. Kaiser, Y. Tasaka, M. Minami **Determination of residual strength parameters of jointed rock masses using *GSI* system** Int. J. Rock Mech. Min. Sci., 44 (2007), pp. 247–265

11. M. Cauvin, T. Verdel, S. Romuald **Modelling uncertainties in mining pillar stability analysis** Risk Anal., 29 (2009), pp. 1371–1380

12. G. Chen, Z. Jia, J. Ke **Probabilistic analysis of underground excavation stability** Int. J. Rock Mech. Min. Sci., 34 (1997), pp. 51.e1–51.e16

13. D.F. Coates **Rock mechanics principles. CANMET, energy, mines and resources Canada** Monograph, 874 (1981)

14. S.H. Dai, M.O. Wang **Reliability Analysis in Engineering Applications** Van Nostrand Reinhold, New York (1992)

15. Deere, D.U., 1968. Geological consideration, rock mechanics in Engineering Practice, Division of Civil Engineering, School of Engineering, University of Wales, Swansea. Wiley, New York, pp. 1–20.

16. J. Deng, Z.Q. Yue, L.G. Tham, H.H. Zhu **Pillar design by combining finite element methods, neural networks and reliability: a case study of Feng Huangshan copper mine, China** Int. J. Rock Mech. Min. Sci., 40 (2003), pp. 585–599

17. P. Dohyun, K. Hyung-Mok, R. Dong-Woo, S. Won-Kyong, S. Choon **Application of a point estimate method to the probability limit-state design of underground structures** Int. J. Rock Mech. Min. Sci., 51 (2012), pp. 97–104

18. G. Exadaktylos, M. Stavropoulou **A specific upscaling theory of rock mass parameters exhibiting spatial variability: analytical relations and computational scheme** Int. J. Rock Mech. Min. Sci., 45 (2008), pp. 1102–1125

19. J. Faraway, C. Chatfield **Time series forecasting with neural network: a comparative study using the airline data** Appl. Stat., 47 (1998), pp. 231–250

20. A.T.C. Goh, F.H. Kulhawy **Neural network approach to model the limit state surface for reliability analysis** Can. Geotech. J., 40 (2003), pp. 1235–1244

21. D.V. Griffiths, G.A. Fenton, C.B. Lemons **Probabilistic analysis of underground pillar stability** Int. J. Numer. Anal. Methods Geomech., 26 (2002), pp. 775–791

22. E. Hoek, E.T. Brown **Practical estimates of rock mass strength** Int. J. Rock Mech. Min. Sci., 34 (1997), pp. 1165–1186

23. E. Hoek, M.S. Diederichs **Empirical estimation of rock mass modulus** Int. J. Rock Mech. Min. Sci., 43 (2006), pp. 203–215

24. E. Hoek, P.K. Kaiser, W.F. Bawden **Support of Underground Excavations in Hard Rock** A.A. Balkema, Rotterdam Brookfield (1995)

25. Hoek, E., Carranza-Torres, C., Corkum, B., 2002. Hoek-Brown failure criterion – 2002 ed. In: Proceedings of the 5th North American Rock Mechanics Symposium and 17th Tunnelling

Association of Canada Conference: NARMS-TAC 2002, July 7–10, University of Toronto, pp. 267–271. <www.rocscience.com> (2.10.02).

26. K. Hornik, M. Stinchcombe, H. White **Multilayer feed-forward networks are universal approximators** Neural Networks, 2 (1989), pp. 359–366

27. M.A. Idris, D. Saiang, E. Nordlund **Probabilistic analysis of open stope stability using numerical modeling** Int. J. Min. Mineral Eng., 3 (2011), pp. 194–219

28. Itasca, 2012. FLAC3D – Fast Langrangian Analysis of Continua in Three – Dimension, Version 5.0. <www.itascacg.com>.

29. Joughin, W.C., Swart, A.H., Wesselo, J., 2000. Risk based chromite pillar design: Part II- Non-linear modelling. In: Proceedings SANIRE Symposium.

30. Kanellopoulas, G.G. Wilkinson **Strategies and best practice for neutral network image classification** Int. J. Remote Sensing, 18 (1997), pp. 711–725

31. K. Kim, H. Gao **Probabilistic approaches to estimating variation in the mechanical properties of rock masses** Int. J. Rock Mech. Min. Sci. Geomech. Abst., 32 (1995), pp. 111–120

32. N. Krauland, P.-E. Söder **Determining pillar strength from pillar failure observation** Eng. Min. J., 8 (1987), pp. 34–40

33. H.Z. Li, B.K. Low **Reliability analysis of circular tunnel under hydrostatic stress field** Comput. Geotech., 37 (2010), pp. 50–58

34. P.A. Lilly, J. Li **Estimating excavation reliability from displacement modeling** Int. J. Rock Mech. Min. Sci., 37 (2000), pp. 1261–1265

35. P.J. Lunder, R. Pakalnis **Determination of the strength of hard-rock mine pillars** Bull. Can. Min. Metall., 90 (1997), pp. 51–55

36. Marinos, P., Hoek, E., 2000. GSI: a geologically friendly tool for rock mass strength estimation. In: Proceedings of the GeoEng2000 at the International Conference on Geotechnical and Geological Engineering, Melbourne, Technomic Publishers, Lancaster, pp. 1422–1446..

37. Marklund, P.I., 2013. Personal Communication.

38. C.D. Martin, W.G. Maybee **The strength of hard-rock pillars** Int. J. Rock Mech. Min. Sci., 37 (2000), pp. 1239–1246

39. MATLAB, 2012. Version 2012b. The MathWorks Inc. <www.mathworks.com>.

40. M. Najafi, S.E. Jalali, A.R. Bafghi, F. Sereshki **Prediction of the confidence interval for stability analysis of chain pillars in coal mines** Saf. Sci., 49 (2011), pp. 651–657

41. L. Obert, W.I. Duvall **Rock Mechanics and the Design of Structures in Rock** John Wiley & Sons Inc., New York (1967) P.A. Oreste **Probabilistic design approach for tunnel supports** Comput. Geotech., 32 (2005), pp. 520–534

42. Palisade Corporation, 2001. @RISK version 4.0. <www.palisade.com>.

43. Palmström, A., 1995. RMi – a rock mass characterization system for rock engineering purposes. Ph.D. Thesis, University of Oslo, Norway.

44. R.J. Pine **Risk analysis design applications in mining geomechanics** Trans. Inst. Min. Metall, Sect. A (1992), pp. A149–A158

45. S.D. Priest, E.T. Brown **Probabilistic stability analysis of variable stopes** Trans. IMM. Sect. A, 92 (1983), pp. A1–A12

46. D. Recio-Gordo, R. Jimenez **A probabilistic extention to the empirical ALPS and ARMPS systems for coal pillar design** Int. J. Rock Mech. Min. Sci., 52 (2012), pp. 181–187

47. L. Rethati **Probabilistic solutions in geotechnics** Developments in Geotechnical Engineering, vol. 46, Elsevier, Amsterdam (1988)

48. E. Rosenblueth **Two-point estimates in probabilities** Appl. Math. Modell., 5 (1981), pp. 329–335

49. D.E. Rumelhart, G.E. Hinton, R.J. Williams **Learning internal representation by error propagation** D.E. Rumelhart, J.L. McClelland (Eds.), Parallel Distributed Processing, MIT press, Cambridge (1986), pp. 318–362

50. M.A. Sahin, M.B. Jaksa, H.R. Maier **Artificial neural network applications in geotechnical engineering** Sakurai, S., 1981. Direct strain evaluation technique in construction of underground opening. In: Proceedings of 22th US Symp. Rock Mech, MIT, Cambridge, pp. 278–282.

51. Sakurai, S., 1997. Strength parameters of rocks determined from back analysis of measured displacements. In: Proceedings of the First Asian Rock Mechanics Symp. ISRM, Seoul, pp. 95–99.

52. Sakurai, S., 1999. Interpretation of field measurements in tunnel practice. In: Proceedings of the Ninth International Congress of Rock Mechanics, Paris, pp. i–vii.

53. H.F. Schweiger, R. Thurner, R. Pöttler **Reliability analysis in geotechnics with deterministic finite elements** Int. J. Geomech., 1 (2001), pp. 389–413

54. Seibi, S.M. Al-Alawi **Prediction of fracture toughness using artificial neural networks (ANNs)** Eng. Fract. Mech., 56 (1997), pp. 311–319

55. Sjöberg, J., 1992. Failure modes and pillar behaviour in the Zinkgruvan mine. In: Proc. 33rd U.S. Rock Mech. Symp (Santa Fe, June 8–10, 1992). Rotterdam: A.A. Balkema, pp. 491–500.

56. Söder, P.E., Krauland, N., 1990. Determination of pillar strength by full scale tests in the Laisvall mine. In: Proceedings of 11th Plenary Scientific Session of the International Bureau Strata Mechanics, World Mining Congress, Novosibirsk, pp. 39–59.

57. H. Sonmez, C. Gokceoglu, H.A. Nefeslioglu, A. Kayabasi **Estimation of rock modulus: for intact rocks with an artificial neural network and for rock masses with a new empirical equation** Int. J. Rock Mech. Min. Sci., 43 (2006), pp. 224–235

58. A.S. Tawadrous, D. DeGagne, M. Pierce, D. Mas Ivars **Prediction of uniaxial compression PFC3D model micro-properties using artificial neural networks** Int. J. Numer. Anal. Meth. Geomech., 33 (2009), pp. 1953–1962

59. J. Torano, R. Rodriguez, P. Ramirez-Oyanguren **Probabilistic analysis of subsidence-induced strains at the surface above steep steam mining** Int. J. Rock Mech. Min. Sci., 37 (2000), pp. 1161–1167

60. U.S. Army Corps of Engineers, 1997. Engineering and design introduction to probability and reliability methods for use in geotechnical engineering. Engineering Technical Letter No. 1110-2-547, Department of the Army, Washington, DC.

61. R.K. Wattimena, S. Kramadibrata, I.D. Sidi, M.A. Aziz **Developing coal pillar stability chart using logistic regression** Int. J. Rock Mech. Min. Sci., 58 (2013), pp. 55–60